Ordnance Survey Mapwork

A course for first examinations

B.D.R. Worthington
formerly Senior Master, Colfe's School, Lee

Robert Gant
Senior Lecturer in Geography, Kingston Polytechnic

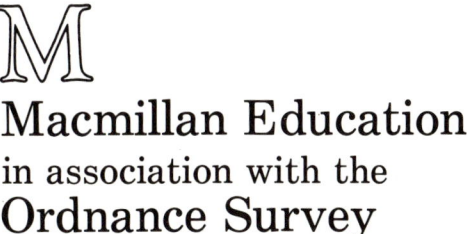

Macmillan Education
in association with the
Ordnance Survey

© B.D.R. Worthington and R. Gant 1983

All rights reserved. No part of this publication may be reproduced or transmitted, in any form or by any means, without permission.

First published 1983
Reprinted 1983

Published by
MACMILLAN EDUCATION LIMITED
Houndmills Basingstoke Hampshire RG21 2XS
and London
Associated companies throughout the world

Printed in Hong Kong

British Library Cataloguing in Publication Data
Worthington, Basil
   Ordnance Survey mapwork.
   1. Maps 2. Great Britain—Maps
   I. Worthington, Basil   II. Gant, Robert
912'.01'4   GA791

ISBN 0-333-30560-4

Contents

List of map extracts   iv

List of aerial photographs   iv

Acknowledgements   iv

Introduction   v

1 · The Brighton Region
- 1A Introduction   1
- 1B Interpretation of the physical landscape   7
- 1C Air photographs and the landscape   12
- 1D Regions in the physical landscape   14

2 · Edale and the Stour Valley
- 2A River study   20
- 2B Model of a river: headwaters to estuary   21
- 2C The drainage system: guidelines for description   28
- 2D Techniques for the comparison of river basins   29

3 · Snowdonia
- 3A Location of a map extract   32
- 3B Landforms in a glaciated upland region   32
- 3C Human activity in an upland environment   38

4 · Wharfedale
- 4A Physical landscape   41
- 4B Human activity in the upland environment   45

5 · South Devon
- 5A Coastal landforms   48
- 5B Settlement and the physical environment   59
- 5C Communications   60

6 · Central Bath
- 6A Physical setting   67
- 6B Historical growth of Bath   68
- 6C Urban environment   71

7 · Lower Teesside
- 7A Physical background   73
- 7B Industrial land use   75
- 7C The process of urban growth   78

8 · South East London
- 8A Introduction   85
- 8B Density of the built-up area   85
- 8C Land use on the urban–rural fringe   94
- 8D Road and rail communications   97

9 · South Yorkshire
- 9A Physical landscape   101
- 9B Economic activity and the rural landscape   102
- 9C The study of rural settlement   104

Index   111

Ordnance Survey Map Extracts

The maps in this publication are reproduced from Ordnance Survey maps with the permission of the Controller of Her Majesty's Stationery Office, Crown copyright reserved.

	Page
The National Grid	3
Symbols for 1:50 000 First Series maps	4
Symbols for 1:50 00 Second Series maps	5
Symbols for 1:25 000 and 1:10 000 maps	6
Map 1 The Brighton Region	23
extract from 1:50 000 Landranger Second Series Sheet 198 (metric values on imperial contours)	
Map 2A Edale	24
extract from 1:25 000 First Series Sheet SK18 (imperial contours)	
Map 2B The Stour Valley	25
extract from 1:50 000 Landranger First Series Sheet 194 (metric values on imperial contours)	
Map 3 Snowdonia	26
extract from 1:50 000 Landranger Second Series Sheet 115 (fully metric)	
Map 4 Wharfedale	51
extract from 1:25 000 First Series Sheet SD97 (imperial contours)	
Map 5 South Devon	52-3
extract from 1:50 000 Landranger Second Series Sheet 202 (fully metric)	
Map 6 Central Bath	54
extract from 1:10 000 Sheets ST76NW, ST76NE, ST76SW, ST76SE (metric values on imperial contours)	
Map 7 Lower Teesside	87
extract from 1:50 000 Landranger Second Series Sheet 93 (fully metric)	
Map 8 South East London	88-9
extract from 1:50 000 Landranger Second Series Sheet 177 (fully metric)	
Map 9 South Yorkshire	90
extract from 1:50 000 Landranger Second Series Sheet 111 (metric values on imperial contours)	

Aerial Photographs

		Page
Plate 1A	The Brighton Region	18
Plate 1B	The Brighton Region	19
Plate 2A	Edale	22
Plate 2B	The Stour Valley	27
Plate 3	Snowdonia	36
Plate 4	Wharfedale	44
Plate 5A	South Devon	56
Plate 5B	South Devon	58
Plate 6	Central Bath	69
Plate 7	Lower Teesside	76
Plate 8	South East London	91
Plate 9	South Yorkshire	108

The authors and publishers wish to thank the following for the use of photographs:

Aerofilms Plates 1A, 6, 7, 8, 9
Airview (M/Cr) Ltd Plates 2A, 4
Cambridge University Collection Plates 1B, 3, 5A, 5B
Crown Copyright Reserved – Ordnance Survey Plate 2B

The publishers have made every effort to trace the copyright holders, but if they have inadvertently overlooked any, they will be pleased to make the necessary arrangements at the earliest opportunity.

Introduction

Several aspects of a student's basic training in geography are tested in CSE, 'O' level and 16-plus examinations. These include visualising the landscape shown on a map (sometimes with the aid of an oblique air photograph); making reasoned correlations between features in the physical and human landscapes; selecting map evidence for the preparation of illustrative sketch maps and diagrams; and, finally, setting out simple arguments and conclusions in writing.

This book aims to meet these examination requirements by developing a structured and logical approach to the interpretation of representative British landscapes from Ordnance Survey maps and related air photographs. Most of the nine maps are metric and all are the latest editions available, at the time of writing, at the scales of 1:50 000, 1:25 000 and 1:10 000.

Each chapter is introduced by a plan which outlines the major themes considered, together with the range of techniques employed. Wide-ranging questions are set to encourage students to think widely and to appreciate the processes which fashion the landscape and the problems of life in the particular environments.

Chapter 1 revises basic mapwork techniques and then considers the techniques for delimiting physical regions in the landscape. In Chapter 2 the role of rivers in the formation of the landscape is described and analysed. Chapters 3 – 5 relate more closely to the interpretation of representative physical environments in Britain, namely a glaciated upland, limestone region and the coastline. Several interrelated themes in the human landscape are covered in Chapters 6 – 9. These include the identification of land-use regions in towns, the effects of urban growth on the countryside, the description and graphical study of communication networks and the analysis of industrial landscapes.

CHAPTER 1
The Brighton Region

Chapter Plan

1A · **Introduction**
    1A.1 Representation of physical and human features
    1A.2 Revision of mapwork techniques

1B · **Interpretation of the physical landscape**
    1B.1 Methods for representing relief
    1B.2 Rock types and characteristic landforms
    1B.3 Techniques for analysing the physical landscape

1C · **Air photographs and the landscape**

1D · **Regions in the physical landscape**
    1D.1 Method of approach
    1D.2 Physical divisions in the Brighton Region

1A · Introduction

1A.1 · Representation of physical and human features

Ordnance Survey maps reproduce a selection of the physical and human features seen in the landscape of Britain. The amount of detail represented, using a range of conventional signs and symbols, is governed by the scale of the Map Series (see pages 4–6). Compare, for example, the local detail and range of symbols used on Map 1 (Brighton Region 1:50 000), Map 4 (Wharfedale 1:25 000) and Map 6 (Central Bath 1:10 000). Examine these map extracts and note the similarities and differences in the methods for representing:

(i) relief and drainage
(ii) vegetation and land use
(iii) settlement
(iv) communications
(v) economic activity

1A.2 · Revision of mapwork techniques

Although you are probably familiar with the map-work techniques described here, to help with later exercises in this book you might find it helpful to refresh your memory.

Map references. These are necessary to locate points on a map. As the map on p.3 shows, within the national grid any point can be defined by a combined value for a northing and an easting.

For convenience this national grid system is divided into 100-kilometre squares which are identified by two letters as shown on page 3. These 100-kilometre squares are further subdivided and are shown on all Ordnance Survey maps at one-kilometre intervals, thus providing a simple means of giving an accurate reference to any place on the map.

To give a reference proceed as follows.
1. Identify the south-west corner of the square in which the place falls.
2. At this intersection read off the grid number of the vertical (eastings) line in the north and south margins and the grid number of the horizontal (northings) line in the east or west margins.
3. The four-figure grid reference of the square in which the place falls is obtained by combining the former with the latter. A more accurate six-figure reference can be obtained by estimating the distance in tenths east and north of the point of intersection.

To make this reference into a unique one for the whole country it should be prefixed by the two letters representing the 100 kilometre square in which the place falls.

Example: Sheet Number 177 East London Metric Second Series. (Map 8 in this book is an extract from this larger Ordnance Survey Map at a scale 1:50 000.) On this sheet the grid reference for the railway station at Crayford is TQ 515745 (Map 8 and Fig. 1.1), as shown overleaf.

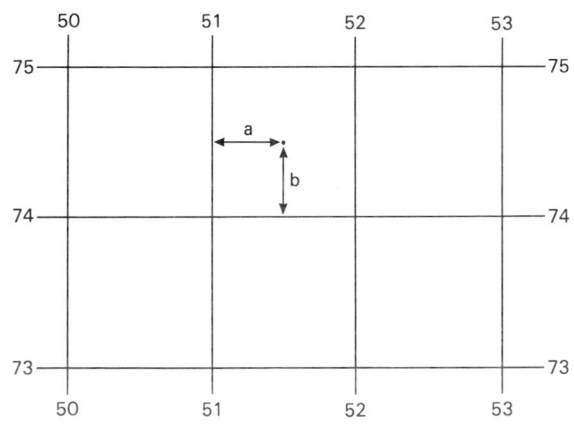

Fig. 1.1 The six-figure map reference

100 kilometre square TQ

Easting 515   Grid number 51 shown in north and south margins and an estimated distance of 5 tenths (see *a* on Fig. 1.1) east of grid line 51.

Northing 745   Grid number 74 shown in east and west margins and an estimated distance of 5 tenths (see *b* on Fig. 1.1) north of grid line 74.

When giving a reference, 'Eastings' should always be given before 'Northings'. (Remember by: 'Indoors first and upstairs second'.)

1.1 Study Map 4 (Wharfedale), Scale 1:25 000, and give a six-figure reference for a bridge, a spring, a church and a barn. Give four-figure references to the squares in which Hawkswick and Starbotton are situated.

1.2 Study Map 8 (S.E. London) and then say what is found at: 478729, 435736, 438709, 547670.

Scale. The scale of a map is the relationship between the distance recorded on the map and the equivalent distance on the ground. This can be shown in one of three ways:
(a) a statement in words e.g. 2 cm represent 1 kilometre;
(b) a line drawing or linear scale:

(Note that the finer subdivisions are always on the left of the scale);
(c) a representative fraction (e.g. 1:50 000 or 1/50 000).

1.3 Study all the maps in this book and discover how many different scales are used.

1.4 Find the distances on the ground represented by (i) 6cm and (ii) 8.5cm where a map has a scale of 1/25 000; 1/50 000; 1/10 000

Straight line distance. This is best measured in centimetres using a rule. This value can then be converted to the corresponding distance on the ground by referring to the map scale.

1.5 What is the straight line distance between Dartford Station 542743 and Sidcup Station 463727 on Map 8?

There are two methods for measuring *distance along a curved line*, for example a road. Firstly a piece of string can be carefully positioned around the curves. This can then be straightened and measured against the linear scale on the map to find the corresponding distance on the ground. Alternatively you may have access to a simple instrument called an opisometer which can be used to measure distance from the map.

1.6 On Map 3, how far is it from the caravan site in grid square 6265 along the A5 to the mountain rescue post in grid square 6460?

Setting or orientating a map. Place a compass on the map and then rotate the map until magnetic north on the map coincides with the north point on the compass.

Bearings. These are always taken clockwise from the north point. For this, a compass with markings between 0° and 360° is used (see Fig. 1.2)

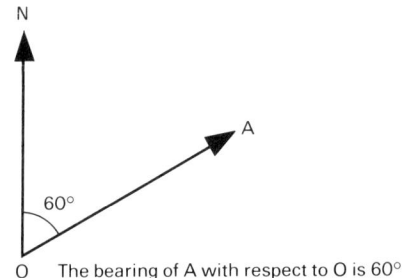

Fig. 1.2 Bearings

1.7 In Map 5, what are the bearings of the following places from the railway station at Kingsbridge: Start Point (8337); Torcross (8242); Bolt Tail (6639)?

Directions can also be given using the points on the compass (Fig. 1.3)

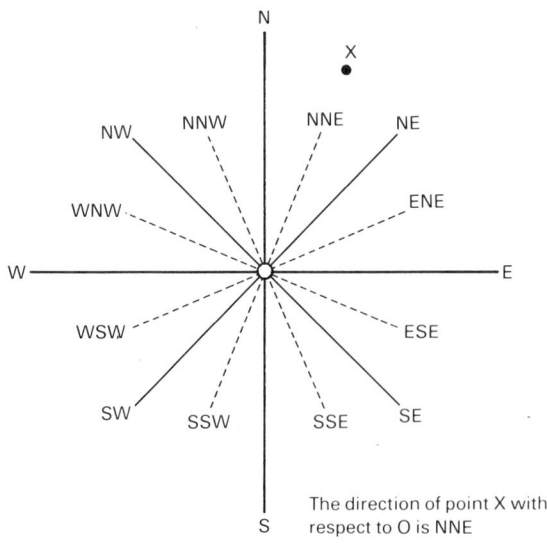

Fig. 1.3 Directions

1.8 What is the direction of Fulking (2411) from Brighton Town Hall on Map 1?

The National Grid

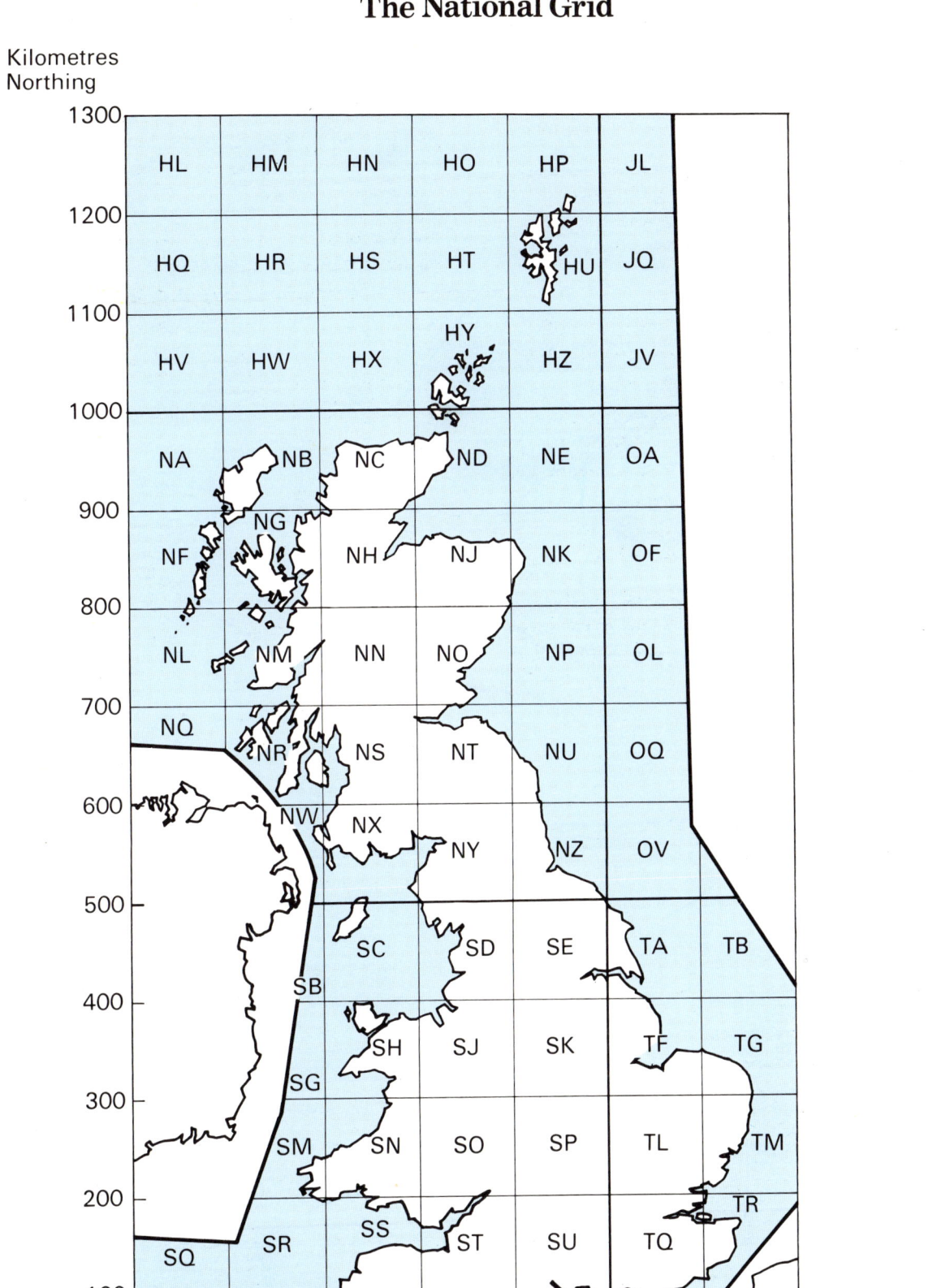

1:50 000 First Series Map
CONVENTIONAL SIGNS

Ordnance Survey

ROADS AND PATHS Not necessarily rights of way

- Service area — M1 — Junction number — Elevated — Motorway
- Motorway under construction
- A 85 (T) — Trunk road
- A 815 — Main road } Single and dual carriageway
- Under construction
- B 845 — Secondary road
- A 558 — B 885 — Narrow road with passing places
- 4·3 metres of metalling or over (not included above)
- Under 4·3 metres of metalling tarred and untarred
- Minor road in towns, drive or track (unmetalled)
- Path
- Gradients: 1 in 5 and steeper   1 in 7 to 1 in 5
- TOLL — Toll gate   Other gates   Entrances to road tunnels
- Road Tunnel

Unfenced roads are shown by short pecks

PUBLIC RIGHTS OF WAY (Not applicable to Scotland)

- Footpath } Public paths
- Bridleway
- Road used as a public path / Byway open to all traffic

Public rights of way indicated by these symbols have been derived from Definitive Maps as amended by later enactments or instruments held by Ordnance Survey on   (date)   and are shown subject to the limitations imposed by the scale of mapping

The representation on this map of any other road, track or path is no evidence of the existence of a right of way

Danger Area   MOD Ranges in the area. Danger! Observe warning notices

WATER FEATURES

- Marsh
- Lake
- Canal and tow path
- Aqueduct
- Ferry P — Ferry (passenger)
- Ferry V — Ferry (vehicle)
- Foot bridge
- Light vessel, lighthouse and beacon
- Highest point to which tides flow
- Slopes
- Cliff
- Flat rock
- Sand and mud
- Sand and shingle
- Low water mark
- High water mark

RAILWAYS

- Multiple } Standard gauge track
- Single
- Narrow gauge
- Mineral line, siding or tramway
- Bridge
- Foot bridge
- Station (a) principal (b) closed to passengers
- Viaduct
- Level crossing
- Tunnel
- Cutting
- Embankment

RELIEF

- —76— Contour values are given to the nearest metre. The vertical interval is, however, 50 feet. If standard metric contours are used, the vertical interval is 10 metres.
- ·144 Heights are to the nearest metre above mean sea level. Heights shown close to a triangulation pillar refer to the station height at ground level and not necessarily to the summit.

1 metre = 3·2808 feet   15·24 metres = 50 feet

GENERAL FEATURES

- Electricity transmission line (with pylons spaced conventionally)
- Pipe line (arrow indicates direction of flow)
- Quarry
- Open pit
- Wood
- Orchard
- Park or ornamental grounds
- Bracken, heath and rough grassland
- Dunes
- Broadcasting station (mast or tower)
- Bus or coach station
- Church } with tower
- or } with spire
- Chapel } without tower or spire
- Glasshouse
- Graticule intersection at 5' intervals
- Triangulation pillar
- Windmill (in use)
- Windmill (disused)
- Wind pump
- Youth hostel

ABBREVIATIONS

P	Post office	TH	Town hall, Guildhall or equivalent
PH	Public house	PC	Public convenience (in rural areas)
CH	Club house	.T	
.MP	Mile post	.A	Telephone call box { PO, AA, RAC }
.MS	Mile stone	.R	

ANTIQUITIES

- VILLA Roman
- Tumulus Non-Roman
- + Site of antiquity
- ⚔ 1066 Battlefield (with date)

BOUNDARIES

- —+—+—+— National
- —○—○—○— London Borough
- (yellow) National Park or Forest Park
- NT / NT } National Trust { always open / opening restricted }
- —·—·—·— County, Region or Islands Area
- —·—·—·— District
- ············ Civil Parish or equivalent

NTS (in red or blue)   National Trust for Scotland

TOURIST INFORMATION

- 🛈 Information centre
- P Parking
- ✕ Picnic site
- ⛺ Camp site
- 🚐 Caravan site
- Selected places of tourist interest
- Viewpoint

1:50 000 Second Series Map
CONVENTIONAL SIGNS

Ordnance Survey

ROADS AND PATHS   Not necessarily rights of way

Symbol	Description
Service area (S), M 1, Junction number 3, Elevated	Motorway (dual carriageway)
M 3	Motorway under construction
Unfenced, A 40 (T), Footbridge, Dual carriageway	Trunk road*
	Main road*
	Main road under construction*
B 284	Secondary Road
A 855, Bridge, B 885	Narrow road with passing places
	Road generally more than 4m wide
	Road generally less than 4m wide
	Other road, drive or track
	Path
	Gradient: 1 in 5 and steeper 1 in 7 to 1 in 5
	Gates Road tunnel
Ferry P, Ferry V	Ferry (passenger) Ferry (vehicle)

* These roads are shown in magenta on Maps 1, 3, 5 & 7, and in red on Maps 8 & 9

PUBLIC RIGHTS OF WAY (Not applicable to Scotland)

- Footpath ⎫ Public paths
- Bridleway ⎭
- Road used as a public path ⎫
- Byway open to all traffic ⎭

Public rights of way indicated by these symbols have been derived from Definitive Maps as amended by later enactments or instruments held by Ordnance Survey on   (date) and are shown subject to the limitations imposed by the scale of mapping

The representation on this map of any other road, track or path is no evidence of the existence of a right of way.

Danger Area   MOD Ranges in the area. Danger! Observe warning notices

WATER FEATURES

- Marsh or salting
- Lake
- Canal, Lock and towpath
- Canal (dry)
- Aqueduct
- Footbridge
- Normal tidal limit
- Lighthouse (in use and disused)
- Beacon
- Slopes
- Cliff
- Flat rock
- Low water mark
- High water mark
- Mud
- Sand
- Dunes
- Shingle

ABBREVIATIONS

P	Post office
PH	Public house
MS •	Milestone MP • Milepost
CH	Clubhouse
PC	Public convenience (in rural areas)
TH	**Town Hall**, Guildhall or equivalent
CG	Coastguard

TOURIST INFORMATION

- Information Centre
- Selected places of tourist interest
- Viewpoint
- Parking
- Picnic site
- Camp site
- Caravan site
- Youth hostel
- Golf course or links
- Bus or coach station
- Public telephone
- Motoring organisation telephone
- Public convenience (in rural areas)

ANTIQUITIES

- VILLA Roman
- Castle Non-Roman
- ⚔ Battlefield (with date)
- ☆ Tumulus
- + Position of antiquity which cannot be drawn to scale
- ₥ Ancient Monuments and Historic Buildings in the care of the Secretaries of State for the Environment, for Scotland and for Wales and that are open to the public

The revision date of archaeological information varies over the sheet

RAILWAYS

- Track multiple or single
- Track narrow gauge
- Freight line, siding or tramway
- Station (a) principal   (b) closed to passengers
- LC Level crossing
- Embankment
- Cutting
- Bridges, Footbridge
- Tunnel
- Viaduct

ROCK FEATURES

outcrop   cliff   scree

HEIGHTS

When standard metric contours are not available the contour interval is 50 ft with values shown to the nearest metre   (Maps 1 & 9)

— 50 — Contours are at 10 metres vertical interval (Maps 3, 5, 7 & 8)

• 144 Heights are to the nearest metre above mean sea level

Heights shown close to a triangulation pillar refer to the station height at ground level and not necessarily to the summit.

1 metre = 3·2808 feet   15·24 metres = 50 feet

GENERAL FEATURES

- Electricity transmission line (with pylons spaced conventionally)
- Pipe line (arrow indicates direction of flow)
- ruin
- Buildings
- Public buildings (selected)
- Quarry
- Spoil heap, refuse tip or dump
- Coniferous wood *
- Non-coniferous wood *
- Mixed wood
- Orchard

* Tree symbols are not shown on Map 9

- Radio or TV mast
- Church ⎫ with tower
- or   ⎬ with spire
- Chapel ⎭ without tower or spire
- Chimney or tower
- Glasshouse
- Graticule intersections at 5' intervals
- Heliport
- Triangulation pillar
- Windmill with or without sails
- Windpump
- Park or ornamental grounds

BOUNDARIES

- National
- London Borough
- District
- County, Region or Islands Area
- National Park or Forest Park
- NT National Trust   always open
- NT National Trust   opening restricted
- NTS (in red or blue) National Trust for Scotland
- FC Forestry Commission   Pedestrians only - observe local signs

1:10 000 SYMBOLS

(Selected as relevant to Map 6 on Page 54)

1:25 000 SYMBOLS

(First Series)

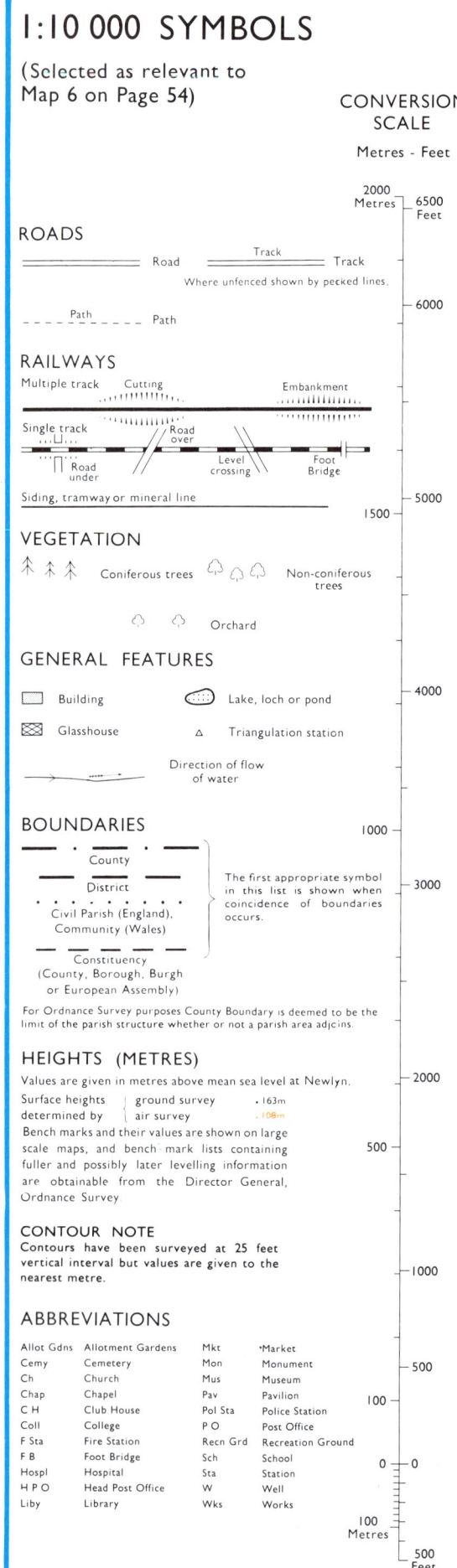

CONVERSION SCALE
Metres - Feet

ROADS

RAILWAYS

VEGETATION

GENERAL FEATURES

BOUNDARIES

For Ordnance Survey purposes County Boundary is deemed to be the limit of the parish structure whether or not a parish area adjoins.

HEIGHTS (METRES)

Values are given in metres above mean sea level at Newlyn.
Surface heights determined by ground survey / air survey
Bench marks and their values are shown on large scale maps, and bench-mark lists containing fuller and possibly later levelling information are obtainable from the Director General, Ordnance Survey.

CONTOUR NOTE

Contours have been surveyed at 25 feet vertical interval but values are given to the nearest metre.

ABBREVIATIONS

Allot Gdns	Allotment Gardens	Mkt	Market
Cemy	Cemetery	Mon	Monument
Ch	Church	Mus	Museum
Chap	Chapel	Pav	Pavilion
C H	Club House	Pol Sta	Police Station
Coll	College	P O	Post Office
F Sta	Fire Station	Recn Grd	Recreation Ground
F B	Foot Bridge	Sch	School
Hospl	Hospital	Sta	Station
H P O	Head Post Office	W	Well
Liby	Library	Wks	Works

Scale 1:25 000

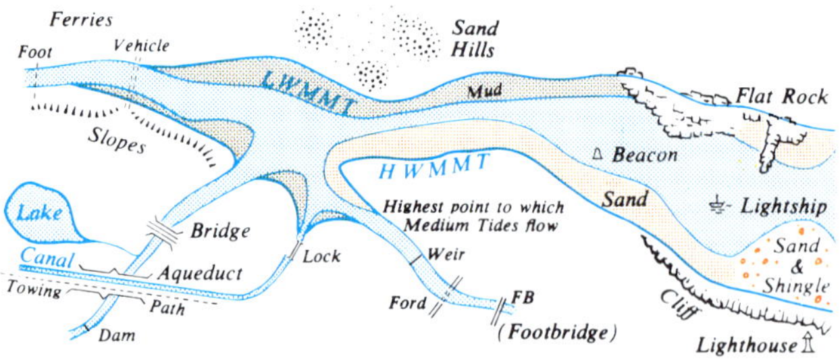

(High & Low Water Mark of Ordinary Spring Tides, in Scotland)

Enlargement and reduction. Fig. 1.4 shows how an area on a map can be increased in scale using a set of grid squares. By doubling the side of each square, the area of the map is increased four times. Similarly, by reducing the side of a square by half, the area is reduced to one quarter.

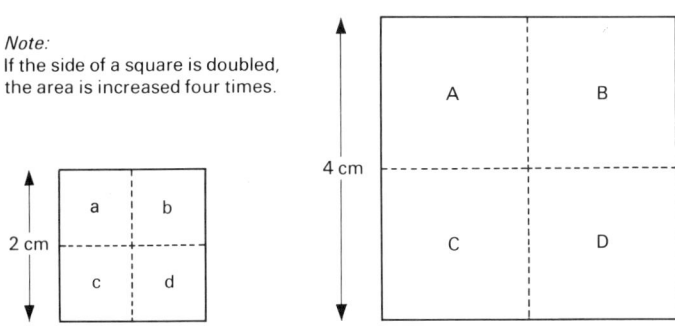

Fig. 1.4 Enlargement and reduction

1.9 Enlarge four times the area of grid square 4970 on Map 8. Insert roads, woods, boundaries, buildings and contours.

Intervisibility between two points (for example O and P on Fig. 1.5) can be measured by constructing a straight line section across the land surface, as explained on pages 10–11. This done, a straight line is drawn on the section to represent the line of sight between the observer (O) and the point being observed (P). If this line of sight cuts the section line, then point P is not visible to the observer.

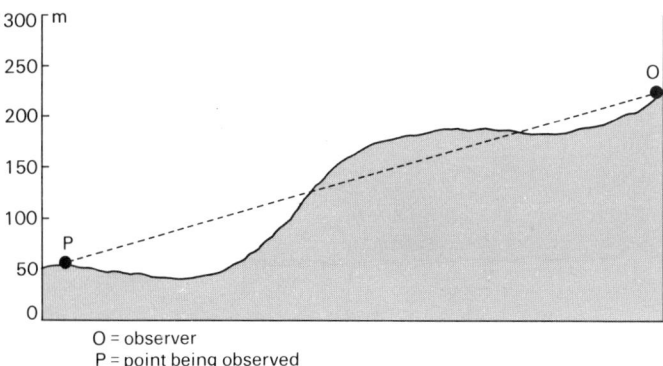

Fig. 1.5 Intervisibility

1.10 On Map 4 can the village of Kettlewell be seen from the Starbotton Road at 968753?

1.11 On Map 6, are the Botanic Gardens in grid square 7365 visible from the school at 742642?

Gradient. The calculation of gradients is fully explained on pages 11–12.

Measurement of area. A method for measuring areas from a map is outlined on page 30.

This study of the Brighton region introduces methods and techniques for the analysis of the physical landscape. It then considers the relationship between land-use patterns and the physical geography of the area.

1B · Interpretation of the physical landscape

1B.1 · Methods for representing relief

The following methods have been used by the Ordnance Survey to represent relief.

Hachures are a series of short lines which look like shading, and were used on the First Edition 1:63 360 maps produced in between 1801 and 1887. These maps are picturesque and give a three-dimensional impression of the landscape. However, with hachures it is not possible to make accurate measurements of the differences in relief from one part of the map to another.

Contours are imaginary lines joining all points of the same altitude above or below a mean sea level datum (Ordnance Datum or O.D.). They are normally continuous across the map and are numbered to indicate their value. In those situations where an abrupt change in the slope of the land surface leads to an excessively close bunching of contours, the lines are broken. Where necessary, too, they are replaced by cliff depiction and complemented by symbols for scree slopes, scars and rock faces.

The interval and colour of contour lines varies with the scale and series of the map.

(i) All maps to a scale of 1:63 360 have contours in brown at an interval of 50 feet. The contour line is drawn more heavily at every interval of 250 feet.

(ii) In 1970 the Ordnance Survey announced the introduction of the 1:50 000 Map Series. Maps in the First Series are photographic enlargements made from the 1:63 360 maps, with contours printed originally in orange and renumbered to the nearest metre. (*Note*: the contour interval remained 50 feet so the irregular sequence of contour values arises from the conversion of feet into metres: 1 metre = 3.2808 feet; 15.2 metres = 50 feet). Hence the metric contour interval is unequal, as shown on Map 1 (Brighton Region

1:50 000) and Map 2B (Stour Valley 1:50 000).

The Second (Landranger) Series 1:50 000 maps currently being produced, show the contours at regular 10 metre intervals, as illustrated on Map 3 (Snowdonia 1:50 000), Map 7 (Lower Teesside) and Map 8 (South East London 1:50 000).

(iii) On the First Series 1:25 000 maps, orange contour lines are drawn at 25 feet intervals (See Map 4, Wharfedale). Second Series (Pathfinder) maps produced to this scale will have a regular contour interval of 5 metres (10 metres in mountain and moorland areas).

(iv) Map 6 (Central Bath) is a composite of metric maps to the scale of 1:10 000. It has the original contours, converted from feet into their nearest metric equivalents. Later revisions of maps at this scale will adopt a regular contour interval of 10 metres.

Triangulation (trigonometrical) pillars and spot heights are shown in black on the 1:63 360 and 1:25 000 First Series maps, and have their heights recorded in feet, but Second Series (Pathfinder) maps are being published using metric heights. Triangulation pillars are drawn in blue on the First and Second Series 1:50 000 maps and like spot heights, have their heights recorded in metres.

1B.2 · Rock types and characteristic landforms

Geological maps provide information on rock types. This can be used to support an interpretation of the physical landscape. In some circumstances, however, where there is no geological information, the nature of the underlying rocks can be deduced from features of the land surface recorded on a topographic map. Certain economic activities associated with particular kinds of rock can provide clues to the underlying geology.

Limestone is probably the most easily identified rock type. In its composition it can range from nearly pure, relatively soft chalk, through less pure Jurassic and older limestones to massive, more resistant, Carboniferous Limestone. Rapid percolation of surface water through the ground occurs on most kinds of limestone. Well-developed joint patterns and excessive percolation can lead to much of the drainage pattern being underground, though in wet weather more of it may flow on the surface.

(i) *Chalk* occurs over large areas of lowland Britain where, generally, it dips at a shallow angle. It normally has beds of sands or clays above and beneath. Its relative resistance to erosion is explained by its permeability rather than its hardness. In contrast to Carboniferous Limestone, chalk supports a more rounded topography. Surface drainage is comparatively sparse or absent, and spring lines occur near the base of the scarp and foot of the dip-slope. There are few natural caves, gorge-like features or scars, owing to the weaker structure of the rock and its dense pattern of joints, compared to Carboniferous Limestone (see below).

Dry valleys, evidence of a former drainage system, are a prominent feature on the dip-slope of the chalk. These valleys are typically long and narrow and are separated by convex spurs (see Fig. 1.6). In chalk regions the historic name for an intermittent stream, 'bourne', is commonly used as the prefix or suffix to a place name, as shown on Map 2B (Stour Valley).

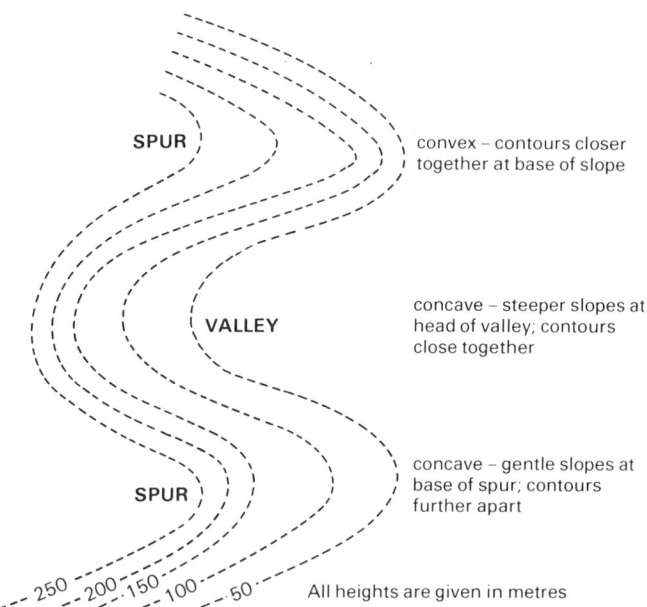

Fig. 1.6 Spurs and dry valleys

Trees are rarely found on chalk and short grass pasture with scrubland is more common on thin soils. However, where superficial deposits of clay-with-flints overlie the chalk, as for example to the north and south of London, deeper and less permeable soils are found. These can support woodland and there may be evidence of surface drainage as the clay-with-flints is impermeable. Areas of ploughed land are also found. Chalk regions may show evidence of prehistoric and historic settlement with, for example, symbols for hill forts, tumuli, Celtic fields and barrows. Concentrations of these monuments would suggest that in earlier times mankind preferred the comparatively open and drier chalk environment to that of the claylands below.

Cement works and lime kilns can indicate a more recent pattern of industrial activity in chalk regions.

(ii) *Carboniferous Limestone* is a massive and more resistant form of limestone found in many parts of upland Britain. It is less pure than chalk and, in consequence, more resistant to erosion. Except in major valleys it supports little surface drainage. This can occur, however, across superficial deposits or beds of impermeable rocks. Maps will normally indicate some features which typify limestone areas. These include caves, pot (swallow) holes, limestone pavements, springs and intermittent drainage. In contrast to chalk, gradients on the valley sides are steeper in limestone areas, and bare rock faces (scars) are sometimes exposed. The alkaline soils are normally treeless, but there may be some cultivation on superficial deposits.

Lime kilns, cement works and quarrying are evidence of industrial activity associated with the limestone. The existence of lead veins in the limestone of the Pennines and Mendips, for example, is suggested on the map by labels for 'old workings' and 'old mine.'

Clay rocks are impermeable and soft and therefore can be easily eroded. They are normally associated with tracts of lowland, having gentle slopes and an abundance of surface drainage. In some areas, artificial drainage is needed to remove surface water. Clay soils are heavy to cultivate and so, before the widespread mechanisation of agriculture, large areas were left under oak woodland. In selected localities brickworks were set up to use local clays.

Sands and sandstones give rise to many different kinds of landscape, depending on their composition. These range from unconsolidated sands to the highly resistant slabs of Millstone Grit found on top of the Pennines.

Less consolidated sandstones may be hard enough to give upstanding relief, but allow the percolation underground of surface water. However, the massive Millstone Grit tends to support surface drainage. In South East England some areas are left as woodland (coniferous) or heathland (gorse, heather, bracken) while others, near large cities, are used for recreation (e.g. golf courses, parks). The higher altitude Millstone Grit moorlands are mainly used as sheep pasture.

Igneous and metamorphic rocks cannot be positively identified without a geological map. Being more resistant than surrounding sedimentary rocks, however, they often form areas of higher land and support surface drainage. Granite can sometimes be detected where a rounded hill topography is associated with prominent, named, tors.

1B.3 · Techniques for analysing the physical landscape

There are several complementary methods for analysing the land surface represented on a map.

Trace of the contour pattern. The nature of the land surface can be appreciated more readily by tracing the pattern of selected contours and adding the altitudes of triangulation pillars and spot heights. The spacing and pattern of contours can then be carefully examined, without the eye being distracted by other information. Figure 1.7 shows the contour pattern on the

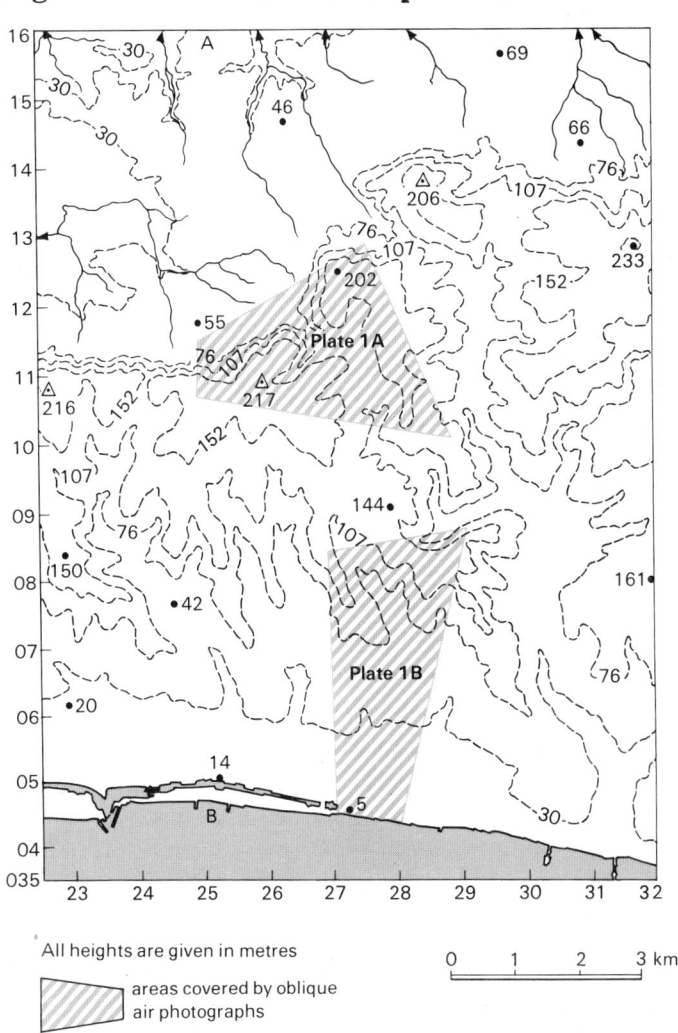

Fig. 1.7 The Brighton region: contour trace

Brighton sheet, together with the main lines of drainage. It identifies the dip-slope in the south, the short scarp facing north and the surface-drained vale beyond.

The contour patterns associated with some typical landforms are shown on Fig. 1.8.

Annotated line sections and sketch sections. Line sections or topographical sections can

All contours are in metres

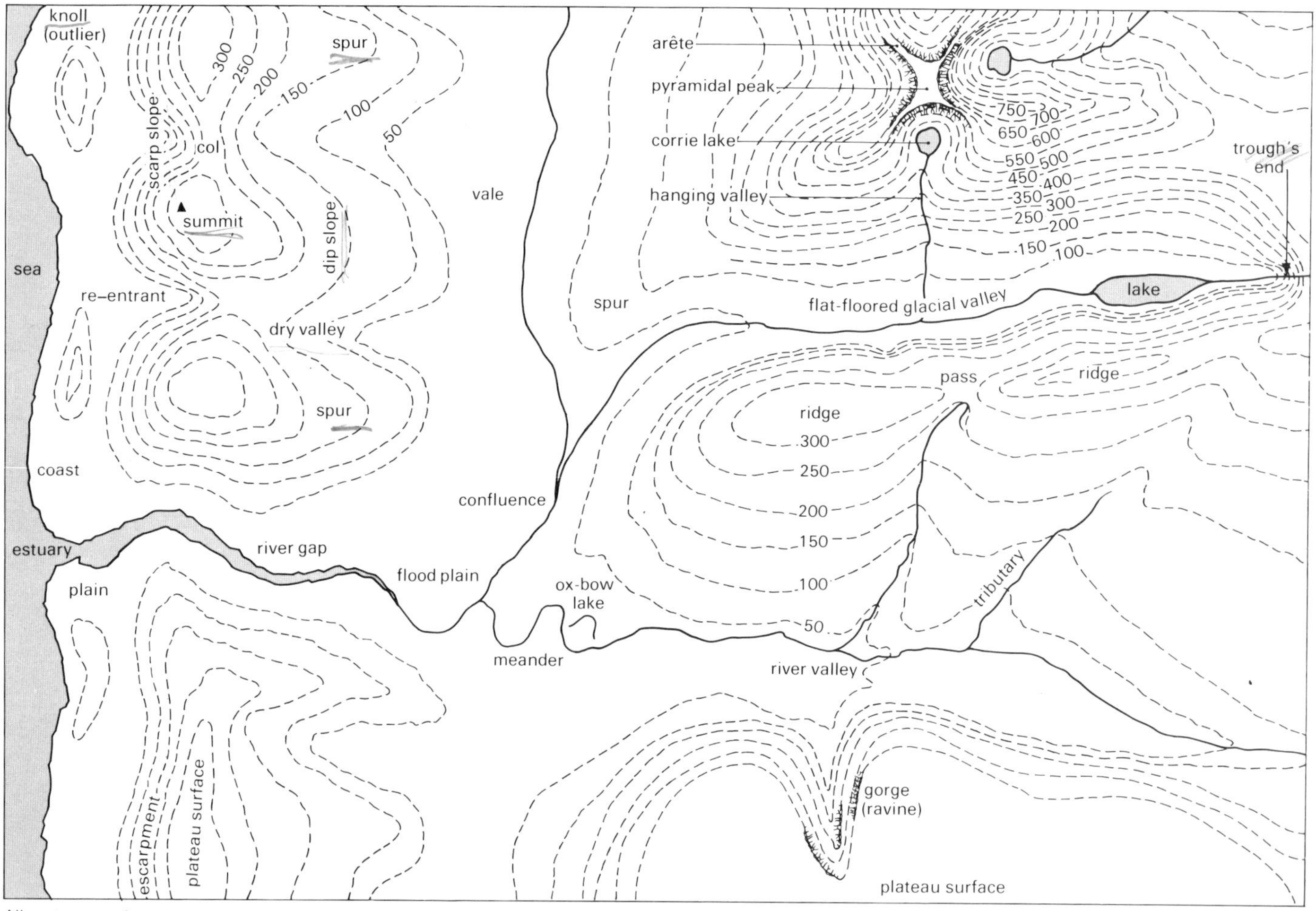

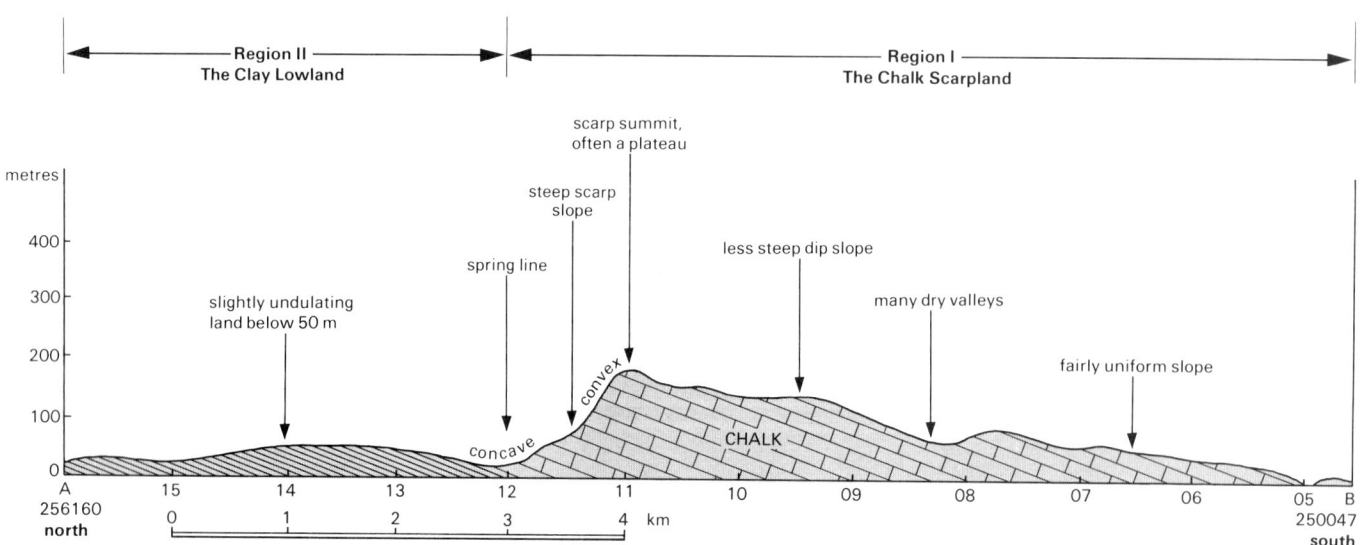

Fig. 1.8 Landscape forms and nomenclature

Fig. 1.9 Line section along Easting 25

provide a very concise and clear summary of the main topographical features in an area. They can be annotated, as in Fig. 1.9, and where possible, information on rock types can be added from a geological map.

Figure 1.9 was constructed along Easting 25 in the following way:

1. A straight-edged strip of paper was placed over the map between points A and B. (See Fig. 1.7)
2. A line of equivalent length was then drawn on graph paper to form the base line for the section. Vertical lines were drawn at each end to indicate altitude. (*Note*: for 1:50 000 maps the vertical exaggeration should not be more than five times; for 1:25 000 maps, not more than four times).
3. From the map the position and value of each contour crossing was noted on the paper strip.
4. The marked strip was then positioned on the graph paper, and the height values marked at appropriate points, perpendicular to the section base-line and using the selected vertical scale.
5. These points were then joined by a smooth line.
6. Significant features in the landscape were indicated by arrows and labelled on the section.
7. Finally, vertical and horizontal scales, a title, direction of section, and necessary grid references were added to complete the diagram.

Sketch sections can be more speedily, though less accurately produced. The framework for the sketch section is constructed as in 1 and 2 above. The position and height of prominent features are then estimated and marked along the section line. The final curved line is then sketched in, all the time noting carefully the steepness of slopes as depicted by the contours on the map. Some considerable practice at drawing sections accurately is required before attempting sketches.

Measurement of slopes
(i) *Gradient.* Figure 1.10 shows the relationship between the spacing of contours and slope in the land surface. Closely-spaced contours depict steeper slopes; widely-spaced contours, more gentle slopes. Compare the contour spacings in grid squares 2411 and 2509 on Map 1. In which grid square does the land slope the more steeply?

The calculation of representative gradients from the map can help with the interpretation of the land surface. Gradient is defined as the average slope of the land between two points. To calculate the gradient, we need to know the distance separating those points on the map (horizontal equivalent = HE) and their difference in height (vertical interval = VI). Figure 1.11 illustrates a gradient calculation made between the spot heights at 251108 and 248093 on Map 1. The gradient of the slope between A and B is

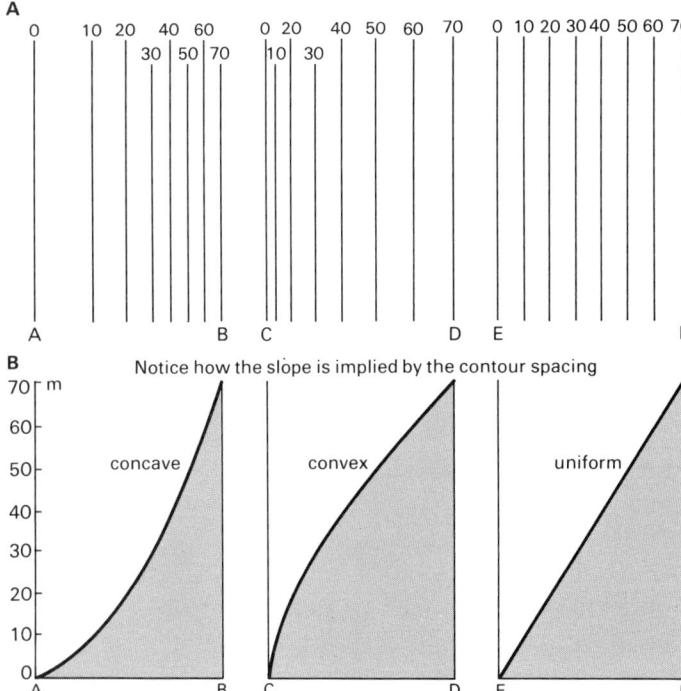

Fig. 1.10 Contour spacing and slope profile

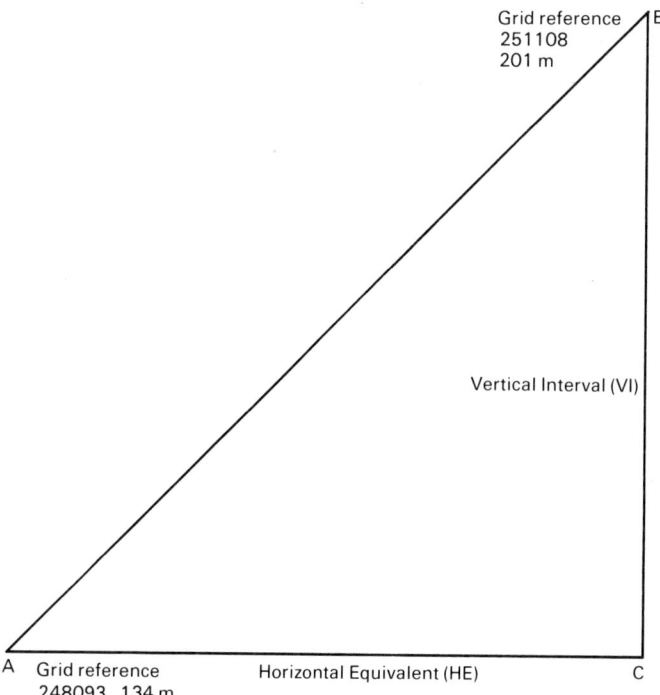

Fig. 1.11 Calculation of gradient

given by the ratio:

$$\frac{VI}{HE} = \frac{BC}{AC}$$

In this case, BC = 201 m − 134 m = 67 metres
AC = 1500 metres.

Therefore, the gradient is $\frac{67}{1500} = \frac{1}{22.4}$ or 1:22.

This ratio can be represented as a percentage, the normal practice:

$1 : 10 = 1$ in $10 = 10\%$ $(\frac{1}{10} \times 100)\%$

$1 : 25 = 1$ in $25 = 4\%$ $(\frac{1}{25} \times 100)\%$

$1 : 22 = 1$ in $22 = 4.6\%$ $(\frac{1}{22} \times 100)\%$

or approximately 4½%.

(ii) *Slope profiles.* From a careful study of the spacing of contours we can appreciate the form or shape of the slopes represented on the map. The terms convex, concave and uniform are often used to describe particular kinds of slope. Examine the arrangement of contours on Fig. 1.10A and relate these to the slope profiles on Fig. 1.10B. Notice that for a uniform slope contours are equally spaced, for a convex one they are closest together at the base, and for a concave one closest at the top of the slope.

Relative relief diagram. In some investigations it is not the general altitude of an area which is important, but the local differences in the height of the land surface. These contrasts can be summarised in a relative relief diagram. This procedure is fully described in Section 4A.2

1.12 Would you find it difficult to ride a bicycle up a hill which has a gradient of 1:20? Would you need to change gear? Draw this slope on graph paper using a scale of one unit on the vertical to twenty units horizontally. Where is the steepest railway track in Britain?

Refer to Map 1 (Brighton Region 1:50 000) and answer the following questions.

1.13 Give a map reference for the highest point shown on the map.

1.14 Which of the statements (i) − (iv) is correct? The land surface in grid square 2509:
 (i) slopes gradually to the north.
 (ii) is flat.
 (iii) slopes gradually and regularly to the south.
 (iv) is undulating.

1.15 What does the V-shape of the contours in grid squares 2508 and 2509 indicate?
 (i) a valley, without a river, sloping south
 (ii) a spur
 (iii) an embankment
 (iv) a plateau

1.16 In which grid square is the steepest slope found?
 (i) 2311
 (ii) 2811
 (iii) 2511
 (iv) 3013

1.17 What is the average gradient of the unclassified road between 276102 (95 m) and 273113 (142 m)?

1.18 Give a six-figure map reference to a point on a road where the gradient is between 1:5 and 1:7. (Look for the steepest slopes.)

1.19 Which statement is correct?
Waterhall (2808) is situated:
 (i) on the summit of a spur which trends south-west to north-east.
 (ii) on a steep east-facing slope.
 (iii) at the bottom of a steep hill.
 (iv) on the floor of a valley which slopes to the east, between 61 m and 76 m.

1.20 Find examples of concave, convex and uniform slopes. Can you explain why there are concave sections at the bottom, and convex sections at the top of so many slopes?

1.21 As a physical geographer, comment on the routes taken by the overhead power transmission line(s) between: (i) 254055 and 243160 (ii) 241080 and 320104.
 What is the relationship between these lines and the network of dry valleys?

1.22 Figure 1.9 shows that Region I comprises the chalk scarp and dip-slope and that Region II is a clay lowland. What map evidence can you assemble to demonstrate the existence of these two rock types? (Refer to Section 1B.2 above). Explain how Plate 1A can help with this exercise.

1.23 Use the procedure described in Section 4A.1 and draw a relative relief diagram covering the area between Eastings 23 and 25, and Northings 08 and 14. What landforms in this region are shown by greater differences in relative relief?

1C · Air photographs and the landscape

Photographs taken from an aircraft can provide additional information to help with the interpretation of the landscape shown on the map. Geographers use two main types of aerial photograph: vertical and oblique. Figure 1.12A shows that vertical photographs are taken from

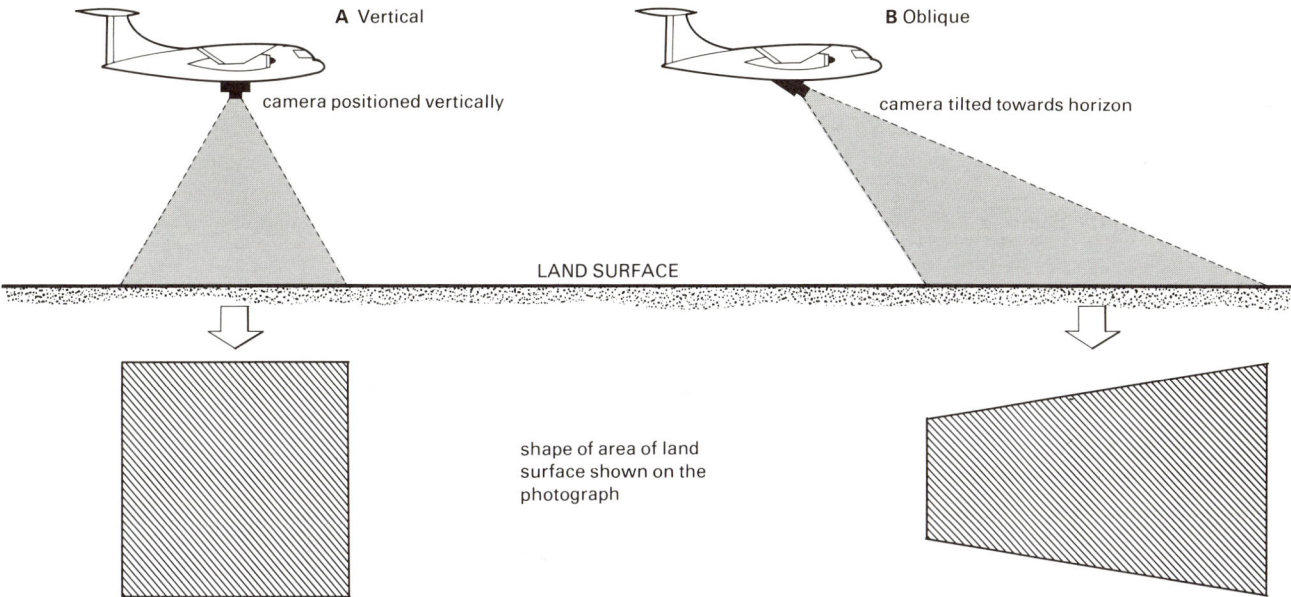

Fig. 1.12 Types of air photograph

the aircraft with the camera pointing directly at the surface of the earth. For oblique photographs the camera is tilted towards the horizon, giving a view of the landscape like that which you would get from an aircraft window or high vantage point such as a mountain peak or tall building (Fig. 1.12B).

1.24 Examine Plates 2B, 5A, 8 and 9. Which of these aerial photographs are (i) oblique (ii) vertical? Give reasons to support your answer.

It is important to recognise two main differences between an Ordnance Survey map and an air photograph.

(i) Ordnance Survey maps, on all scales, use symbols to represent *a selection of features* from the landscape. An aerial photograph, in contrast, records *all visible features* in that landscape.

(ii) The scale of an Ordnance Survey map remains true in all directions. Although we do not need to consider the technical details here, we must be aware that the scale varies on different parts of a photographic print. While vertical photographs relate to rectangular areas, obliques normally cover wedge-shaped tracts of the land surface (see for example, Figures 1.7, 3.10, 4.1, 6.3). On oblique photographs, features in the foreground appear larger than similar features in the background. This means that the scale of the foreground is larger than that of the background. Furthermore, as Fig. 1.13 shows, an abrupt change in the height of the land surface covered by a single print can give rise to marked differences in scale, and therefore detail, from one part of the photograph to another. This feature is clearly demonstrated on Plate 1A, which includes both the top and the bottom of the scarp face.

While we can use our experience to identify many of the features shown on an oblique air photograph, a little more time and practice may be needed to work as easily with vertical air photographs. The following guidelines can be used to interpret a vertical air photograph:

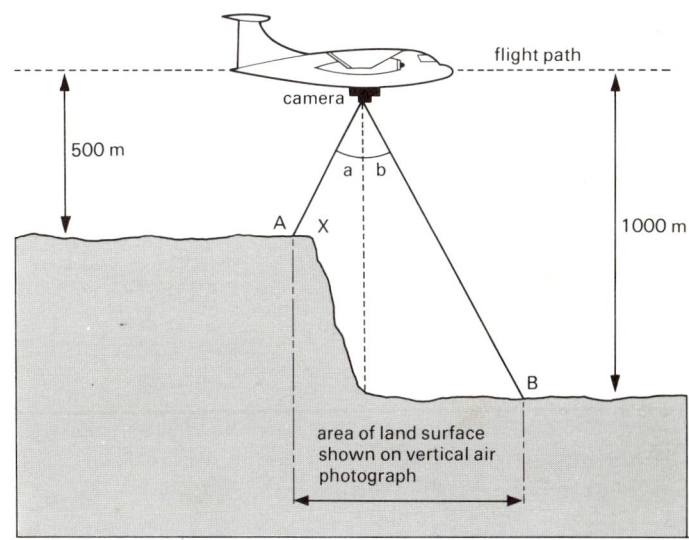

Note: angle a = angle b

The land surface A–X is approximately 500 m higher than X–B. Therefore the scale of features shown on the photographic print will be larger between points A and X than between points X and B. Can you explain why this is so?

Fig. 1.13 Variation in the scale of a vertical air photograph

1. Examine the whole print and identify the main kinds of landscape that are reproduced.

2. Try to recognise individual features. For those objects you cannot readily recognise, examine their size, shape, position, distribution pattern and association with other objects you have successfully identified. Remember that in some instances the cast of a shadow or variation in colour tone on the print can provide additional clues. A hand magnifying glass can be a useful aid for detailed work of this kind.

1D · Regions in the physical landscape

ID.1 · Method of approach

The procedure for delimiting regions in the physical landscape can be divided into three stages.

1. *Decide what features are to be used as the basis for the exercise.* Altitude, the characteristics of the physical land surface, and the drainage pattern are commonly used to distinguish one physical region from another.

2. *Analyse the relief pattern shown on the map.* To organise your thoughts, examine the map closely and prepare answers to the following questions. The techniques outlined in 1B are invaluable in this exercise.

(i) Where are the main areas of 'higher' and 'lower' land? (Note the difference in height between the highest and lowest points shown on the map.)

(ii) What proportion of the map do these areas cover? (Estimate the proportions, and comment on the shape of the areas.)

(iii) Is it possible to draw a boundary line separating these regions on a sketch map?

(iv) Does each region have a distinctive pattern of drainage? (Examine and describe the network, and identify artificial drainage systems.)

(v) What are the characteristic features of the land surface in each of these major regions? (Consider the following: gradients; dissection; relative differences in height.)

(vi) Examine, in turn, each major physical region. Is there a sufficient variation within each region

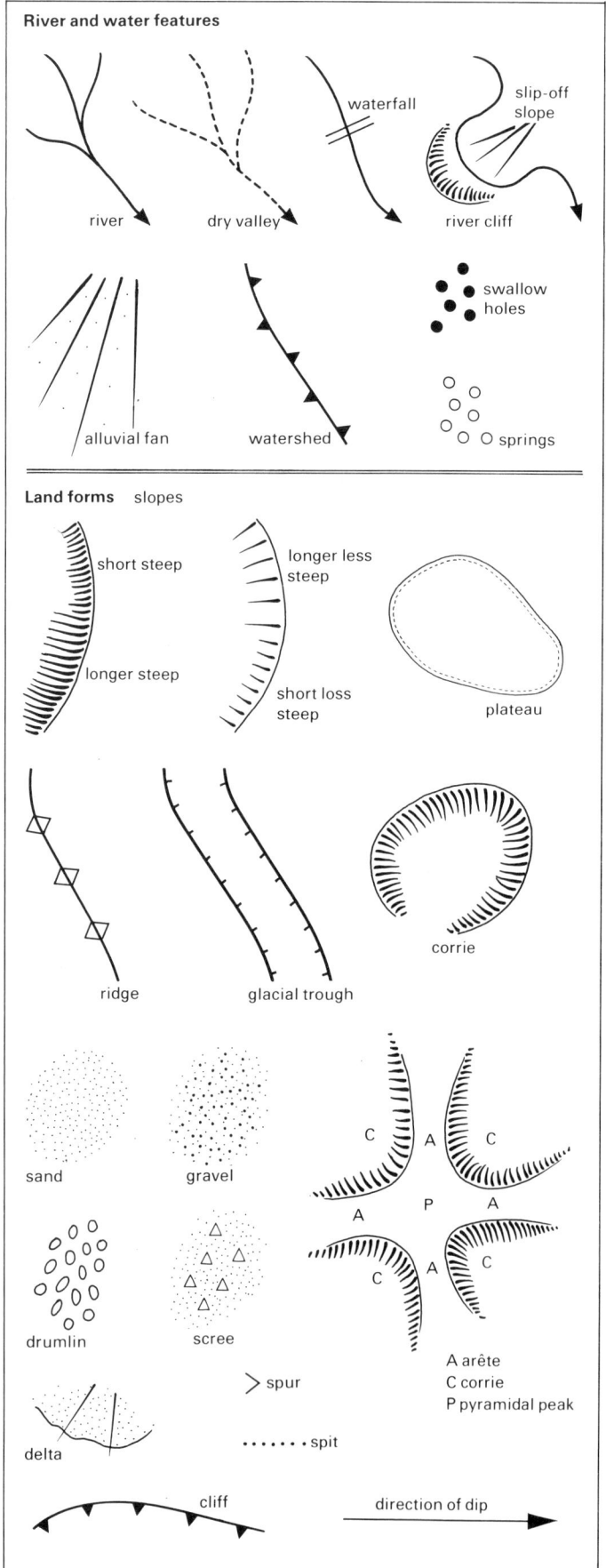

Fig. 1.14 Symbols for use on sketch maps

to justify its further division into sub-regions? (Refer to points (i) – (iv) above.)

3. *Summarise your conclusions on a sketch map.* Prepare a simplified sketch map to show the main physical regions. Use symbols such as those shown in Fig. 1.14 to indicate the principal physical characteristics of each region. Label significant features in each region (e.g. plateau, scarp-face, dip-slope, ridge, watershed, etc) and add brief descriptive notes. Prepare a short written account to justify the set of regions you have proposed. Support each statement with references to map evidence, diagrams, line sections or calculated gradients. Finally, review the problems you experienced in drawing the boundaries between neighbouring regions: for example, were the boundaries clear-cut or did you insert a generalised boundary line?

ID.2 · Physical divisions in the Brighton Region

Delimitation of physical regions. From a careful examination of Map 1, using a variety of techniques, it can be argued that there are two main regions in the physical landscape. These are shown on Fig. 1.15.

Region I comprises a block of relatively higher land and trends east to west across the southern two-thirds of the map. As Figs 1.9 and 1.15 demonstrate, this chalkland region can be divided into two sub-regions: a north-facing and steeper scarp-slope where gradients can approach 1:2 (2311), and a south-facing dip-slope, extending to the coast, with slopes averaging 1:23 (2508, 2509, 2510).

The scarp face extends eastwards for approximately 4 km from grid square 2211. It

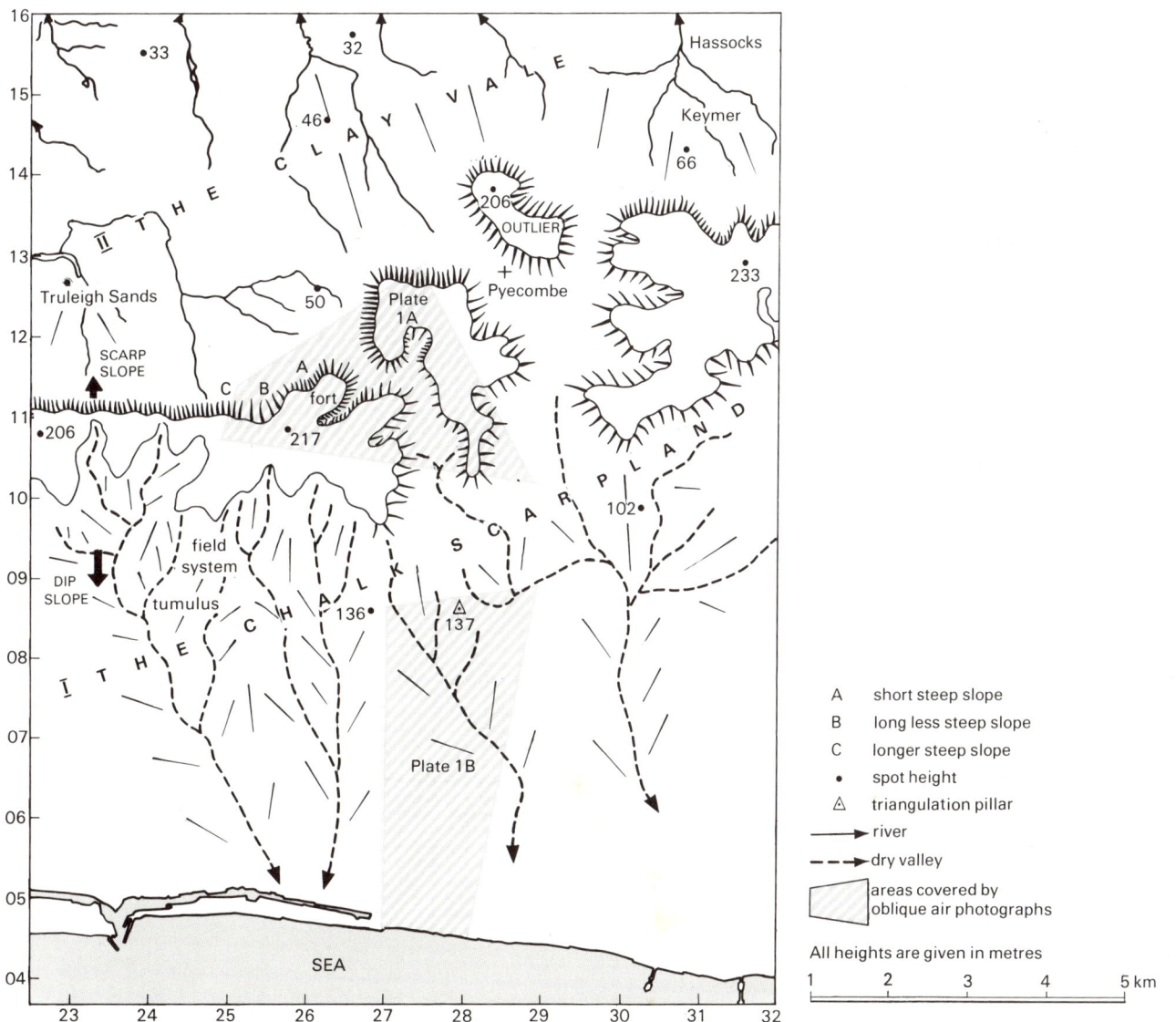

Fig. 1.15 Physical divisions in the Brighton Region

then turns abruptly to the north east at Newtimber Hill (2712). Thereafter, it resumes a general easterly trend for a further 5 km. Several major embayments or re-entrants break the continuity of the scarp face, the two largest occurring at Devil's Dyke (2611) and Pyecombe (2812). Less significant re-entrants also occur as, for example, at 2913 and 3113. Towards the east the wider spacing between the contours indicates that the scarp face is becoming less steep. It is relevant to note that the routes taken by the A23(T) and A273 follow gentler gradients to either side of Wolstonbury Hill (2813). As Fig. 1.9 illustrates, in section the slope of the scarp face is typically convex at the top, concave at the base, and uniform in between.

The altitude of the dip-slope falls from over 200 m at the crest of the scarp to sea level. It is dissected by numerous, branching, dry-valley networks, for example that used by the railway and A23(T) from Pyecombe to the coast. The interfluves, which are those areas of higher land separating neighbouring valleys, are gently rounded. Nowhere is there evidence of surface drainage.

Region II is low-lying. The pattern of contours and spot heights, and the flow direction of small streams, indicate that the land slopes to the north west from an altitude of 60–70 m at the base of the scarp (2411 and 2613) to 30–40 m near Woodmancote (2315). Within this clay vale there are slight undulations in the land surface. Moreover, it may be significant to note that two low mounds which are enclosed by the 30 m contour at Truleigh Sands (2212) and Perching Sands Farm (2412) contain the place name element 'sands'. This may possibly indicate the existence of local sand deposits overlying the clay.

Physical regions and land-use patterns. The exercises in Section 1A confirm that the amount and variety of land-use information given on an Ordnance Survey map vary with its scale and series. It is clear, too, that certain kinds of land use are not recorded on a 1 : 50 000 map. In particular, there is little positive information on the types of agricultural land use.

1.25 What relationship exists between land-use patterns and the major physical regions shown on Fig. 1.15?

This question will be answered by considering two main types of land use: (i) settlement (including industry) (ii) agriculture and natural vegetation.

Firstly, however, you may wish to refer to pages 4 and 5 to refresh your memory on the range of land-use symbols used on 1:50 000 maps.

(i) *Settlement distribution.* Study the distribution of settlement on Map 1, and answer these questions.

1.26 Show on a tracing overlay of the map the distribution of historical settlement. Mark all tumuli, field systems, forts and other historical features such as moats, motte and bailey, Roman remains. Relate this distribution pattern to the physical regions shown on Fig. 1.15. What conclusions can you draw from this exercise?

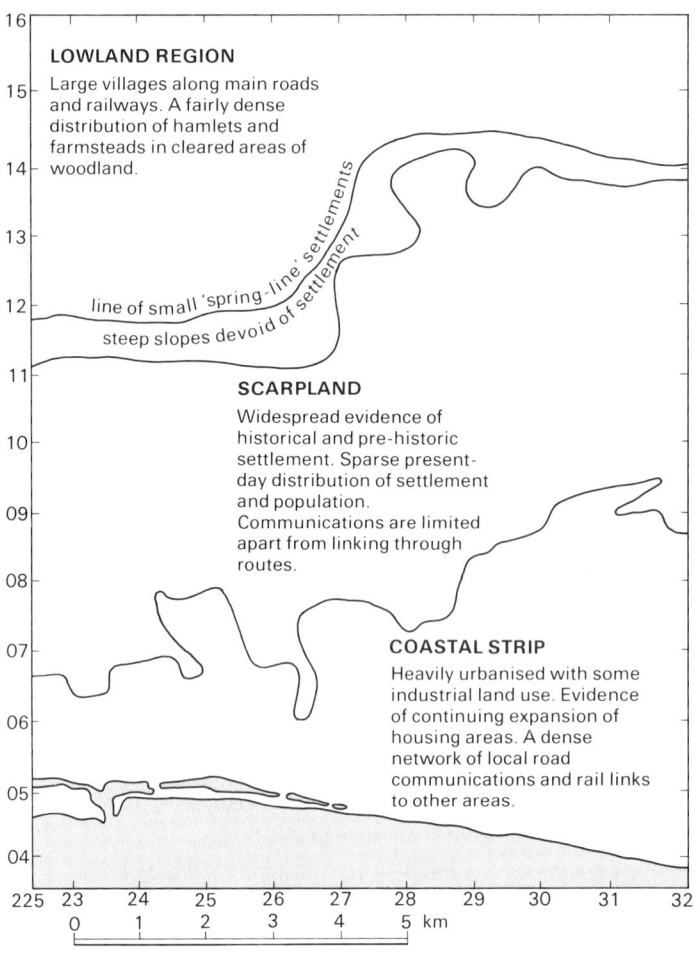

Fig. 1.16 Regions of settlement

1.27 Do you agree with the regional divisions in the settlement pattern proposed on Fig. 1.16? Give reasons to support your answer. Refer to Fig. 1.15, and comment on the relationship between the pattern of settlement and physical geography of the area shown on Map 1. When preparing your answer, note that some settlement may not be obviously related to physical features. Consider, for example, the urbanised coastal settlement. How is it related?

(ii) *Farming activity and physical environment.* What evidence of farming activity is given on an Ordnance Survey map? In general, cropland and pasture are shown in white. Some specialised kinds of agriculture such as orchards and glasshouses have distinctive symbols. In some circumstances, too, the density of named farms may indicate the relative fertility of the land. The larger the scale, the more information can be given, but nevertheless, on the basis of this evidence alone, it would be unwise to specify the detailed nature of farming activity in a region.

Some general guidance on possible kinds of farming activity, however, can be gained by studying the physical environment. This allows us to determine the balance between factors which are broadly favourable and unfavourable to farming activity. The following guidelines can help in this assessment:

1. Determine the main regions in the physical landscape using the procedure outlined in 1D.1.
2. Consider each physical region in turn with regard to the following features which may influence the pattern of farming activity:

(a) *Altitude/relief*: height of land above sea level (above 250 m the climate tends to be too severe for widespread arable cultivation); signs of glaciation which could have removed soil from valley sides; evidence of rock exposures, screes, cliffs. All of these would hinder cultivation of the land.

(b) *Surface characteristics*: variations in gradient; availability of shelter for livestock; aspect (which refers to the direction in which the slope faces). Slopes facing the sun throughout much of the day have climatic advantages for many kinds of agriculture. In contrast, those exposed to the north and north-easterly winds in winter are severely disadvantaged.

(c) *Drainage*: density of natural drainage pattern; artificial drainage; flood hazards; dykes; embankments; straightening of river courses. Severely waterlogged soils cannot be used for arable cultivation and normally only provide summer grazing for livestock.

(d) *Climate*: the location of the area in Britain and its temperature and rainfall regimes.

(e) *Vegetation*: symbols for rough grazing (on some maps), marshland; tree shelter-belts for livestock – direction and orientation; woodland distribution – type (on some maps), pattern and size of blocks.

(f) *Communications*: accessibility of named farms to metalled road surfaces; road type; planned landscapes (e.g. The Fens); relationship between farms and the local road network.

(iii) *Proposed land-use regions.* Region I, labelled as the Chalk Scarpland on Fig. 1.15, can be sub-divided into three land-use regions, as shown on Fig. 1.17. Plate 1A shows that *Region IA*, the steep, north-facing, scarp-face is highly unsuited to arable farming or settlement. It is probably used as sheep pasture.

Region IB, the south-facing, slacker, dip-slope is sparsely populated. However, this region of undulating chalk downland probably supports a

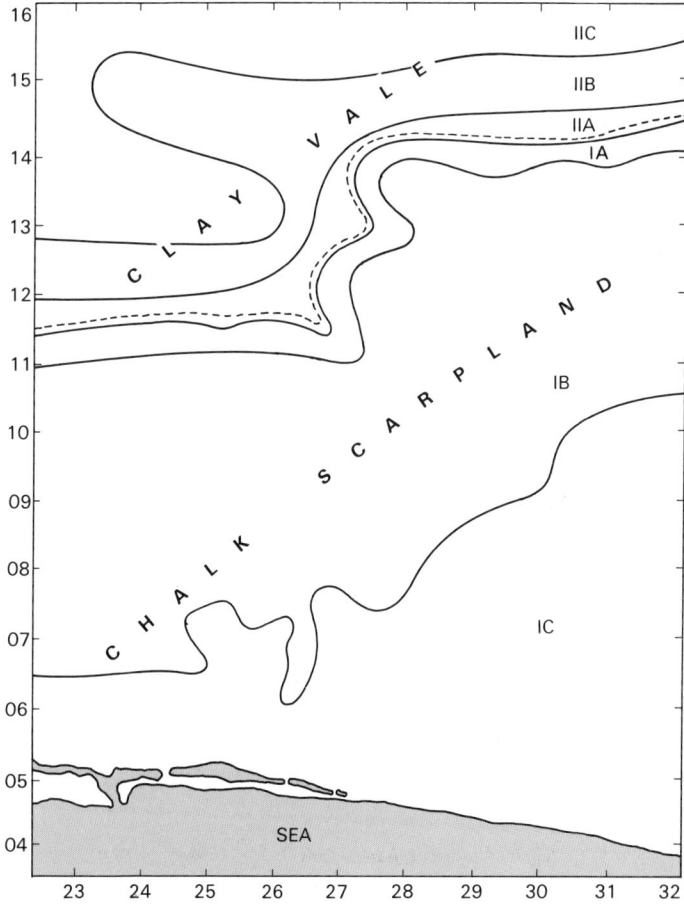

----- boundary between the chalk scarpland and the clay vale

Physical region
I Chalk Scarpland

II Clay Vale

Land-use region
IA sheep pasture
IB sheep/dairying/horticulture
IC urban and industrial

IIA arable
IIB woodland and livestock (cattle)
IIC livestock (cattle)

Fig. 1.17 The relationship between land use and physical regions

similar pattern of sheep farming and cattle rearing especially at higher altitudes. Maybe, too, there are some dairy farms and horticulture units near the urban market in the south. There is, however, no direct evidence on the map to support these statements on farming activity.

Across the dip-slope there is abundant evidence of historical and prehistoric settlement, namely forts (229085, 262112), tumuli (2510), and ancient field systems (2308, 3010). On the fringe of Region IC the landscape is laced with footpaths. The five golf courses, too, relate to the needs of an urban population for outdoor recreation (2607; 2608; 2809). *Region IC* comprises a dense, coastal, ribbon of settlement, communication networks and industrial land uses, as shown on Plate 1B.

Region II lies north of the scarp-face and can be subdivided into three areas. *Region IIA* occupies the lower slopes of the escarpment. It comprises a belt of mainly arable farmland. This is suggested by the number of farms and undulating terrain. It ranges in width from 500 m to 1 km (see Plate 1A). In places, it is separated from the scarp-face by a fragmented, narrow ribbon of trees. This is the spring-line

Plate 1A

Plate 1B

region and there are several villages and farms. They were probably located here initially because of the availability of water and a sheltered position. Poynings (2611) may also have been influenced by a route gap in the escarpment.

Region IIB and *Region IIC* coincide with an area of sands and clays. The surface drainage flows to the north west. In Region IIB the landscape is heavily wooded. Indentations in the perimeters of some woodlands (e.g. Shaves Wood 2514; Parkwood 2613) suggest that either part of the original woodland has been felled, or that additional planting has taken place to provide shelter for livestock. Region IIC has a greater density of farms than Region IIB. The farming activity, however, is probably very similar. There is far less woodland. The recurrence in local place names of the element 'sands' (2212; 2412) might indicate that this area is underlain by sands and not clay. Being better drained and of lighter soil, it is probably more suitable for cultivation than the heavier clays.

1.28 Examine Plate 1A carefully and answer these questions.
 (i) Name the village A, hamlet B and building C. For each, give a map reference.
 (ii) What form of land use occurs at D on the air photograph? Give a map reference for this activity.
(iii) Explain how the air photograph can help in distinguishing the land-use patterns in Regions I and II, shown on Fig. 1.15.
(iv) Are the fields on either side of the minor road between 250115 and 264120 used for different purposes? If so, what differences in land use do you detect? Suggest reasons for this contrast in land use.

1.29 On tracing paper, mark those woodlands greater than 0.1 km² in area. Which physical region on Fig. 1.15 has (i) most woodland; (ii) least woodland?

CHAPTER 2
Edale and the Stour Valley

Chapter Plan

2A · **River study**
  2A.1 Surface water
  2A.2 Human adjustment to rivers

2B · **Model of a river: headwaters to estuary**
  2B.1 Upper course
  2B.2 Middle course
  2B.3 Lower course

2C · **The drainage system: guidelines for description**
  2C.1 Course of a stream/river
  2C.2 Nature of a valley
  2C.3 Drainage networks

2D · **Techniques for the comparison of river basins**
  2D.1 Area of a drainage basin
  2D.2 Drainage density
  2D.3 Average gradient

2A · River study

2A.1 · Surface water

Drainage basins are fundamental units in the study of physical geography. In non-glaciated areas, surface water has been an important influence on the shape of the contours shown on an Ordnance Survey map. Four important factors control the volume of water in streams and rivers, the rate at which it flows, and the amount of erosion and deposition it achieves.

(i) *Gradient*. Water in drainage channels with steep gradients will flow, in general, more quickly than that in gently graded channels.

(ii) *Underlying geology*. Limestone allows rapid percolation; clay will not. (See Section 1B.2). Therefore, limestone areas have less surface drainage than clay areas. This contrast can be determined visually from the map or by calculating statistics for drainage density (see below, Section 2D.2).

(iii) *Vegetation cover*. The runoff of surface water is greater where there is little or no vegetation. Tall, dense vegetation invariably retards the movement of surface water.

(iv) *Aspect*. South-facing slopes are warmer and evaporation is consequently greater.

2A.2 · Human adjustment to rivers

In many situations man has modified the natural drainage channels and, consequently, the flow of water. He has used the water for domestic, industrial, agricultural and recreational purposes. Study Maps 1–9 and answer the following questions.

2.1 Did streams have an influence on the siting of the villages situated at the foot of the scarp on Map 1 (Brighton Region)?

2.2 Examine grid square 7066 on Map 3 (Snowdonia).
  (i) Describe the site of the Dulyn Reservoir.
  (ii) Is there any other evidence of water storage on the map?
  (iii) What factors are important for siting a reservoir?

2.3 Study Map 8 (South East London) and read Chapter 8, Section 8B.2.
  (i) What is the intensive agricultural land use shown in grid squares 4870 and 5169? Why does this require special water supplies?
  (ii) Search the valleys of the River Cray and River Darent for evidence to show that man has modified the flow of these rivers. Explain why these changes were necessary.
  (iii) What evidence is there to support the statement that 'The River Thames and its tributaries have been important in the location of industry'?

2.4 'Rivers can prove a hindrance to communications'. Illustrate this statement from Map 5 and Plate 5A (South Devon); Map 6 and Plate 6 (Central Bath); Map 7 and Plate 7 (Lower Teesside); and Map 8 (South East London).

2.5 Explain why the historical cores of Crayford and Dartford are situated in grid squares 5174 and 5473/5474, respectively, on Map 8 (South East London).

2.6 Use Map 7 (Lower Teesside) and Map 8 (South East London) to explain how you can identify areas of land that are naturally ill-drained.

2.7 'In urbanised areas the system of natural drainage has been widely disrupted by residential and industrial building'. Present evidence from Map 7 (Lower Teesside), Map 8 (South East London) and Map 9 (South Yorkshire) to support this statement.

2.8 What are the attractions of the following river valleys for tourists?
  (i) Nant Ffrancon (Map 3)

(ii) Kingsbridge Estuary (Map 5)
(iii) Wharfedale (Map 4)
(iv) Darent Valley (Map 8)
Support your answer with appropriate evidence from Plates 3, 4 and 5A.

2.9 Use the guidelines on settlement site, morphology and function provided in Chapter 9, 9C.2, to describe the village of Iwerne Minster (8614) on Map 2B. Prepare a sketch map to illustrate your answer.

2B · Model of a river: headwaters to estuary

Figure 2.1 summarises typical changes which occur in the course of a stream and the shape of its valley, as it flows from its source to base level in the sea or a lake. Although this model is greatly simplified, it nevertheless describes the main features of a drainage system which can normally be identified from an Ordnance Survey map.

2B.1 · Upper course

Plate 2A shows part of the valley of the Grinds Brook. Refer to the map and state in which direction the camera was pointing. Follow the course of the Grinds Brook from its source, just west of 102874, to its confluence with the River Noe at 127852. Now answer these questions.

2.10 Refer to Plate 2A and Map 2A.
(i) Name the streams A, B and C.
(ii) Describe the main features of the drainage pattern shown on Plate 2A.
(iii) Study Fig. 2.1. What evidence can you draw from Plate 2A to show that the Grinds Brook is in its upper course?
(iv) Locate Grindslow House (122864), Lands Barn (126862), and Grindsbrook Booth on Plate 2A.

2.11 Draw a longitudinal section down the valley floor (not along the course of the stream). How does the gradient of the valley change? Why does this happen?

2.12 Draw line sections across the valley at three equidistant points between the source and the confluence with the River Noe. How do these sections vary? Suggest reasons for this.

2.13 Count the number of contours which cross the brook in:
(i) one kilometre below 110873
(ii) one kilometre below the ford at 119868.
Explain the difference in your answers.

2B.2 · Middle course

Now study the valley of the River Stour on Map 2B (Stour Valley).

2.14 How many contour lines cross the River Stour?
2.15 In which direction does the river flow?
2.16 What is the average gradient of the valley floor?
2.17 Find examples of a meander, a floodplain and a river cliff. Draw a diagram to illustrate the characteristics of each landform.

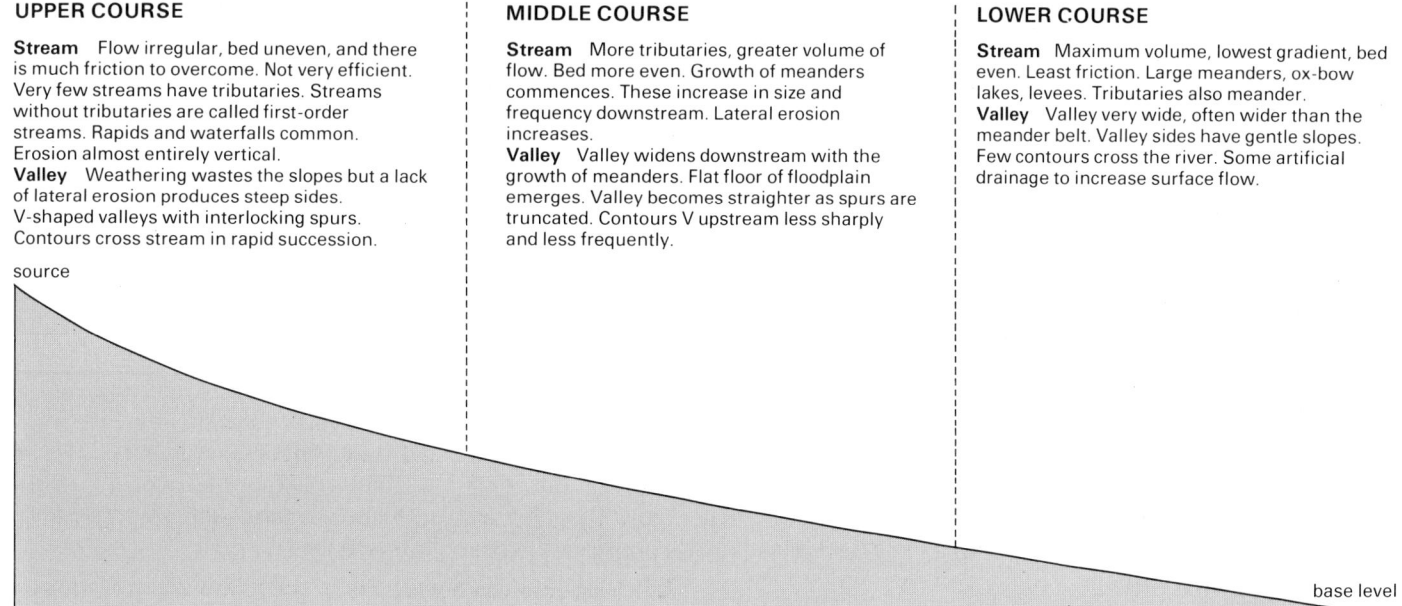

UPPER COURSE

Stream Flow irregular, bed uneven, and there is much friction to overcome. Not very efficient. Very few streams have tributaries. Streams without tributaries are called first-order streams. Rapids and waterfalls common. Erosion almost entirely vertical.
Valley Weathering wastes the slopes but a lack of lateral erosion produces steep sides. V-shaped valleys with interlocking spurs. Contours cross stream in rapid succession.

MIDDLE COURSE

Stream More tributaries, greater volume of flow. Bed more even. Growth of meanders commences. These increase in size and frequency downstream. Lateral erosion increases.
Valley Valley widens downstream with the growth of meanders. Flat floor of floodplain emerges. Valley becomes straighter as spurs are truncated. Contours V upstream less sharply and less frequently.

LOWER COURSE

Stream Maximum volume, lowest gradient, bed even. Least friction. Large meanders, ox-bow lakes, levees. Tributaries also meander.
Valley Valley very wide, often wider than the meander belt. Valley sides have gentle slopes. Few contours cross the river. Some artificial drainage to increase surface flow.

Fig. 2.1 Model of the changes in the character of a stream and its valley from the headwaters to the sea

Plate 2A

2.18 Examine Plate 2A and Plate 2B. Which of these is a vertical air photograph? (See Chapter 1, page 13.) Which kind of photograph, oblique or vertical, is more useful in illustrating features of a river valley? Support your answer with reasons.

2.19 Give map references to indicate the area of the map covered by Plate 2B.

2.20 Identify the river cliff (8706) on Plate 2B. Explain how this feature was formed.

2.21 Compare the following cross sections:
 (i) The Blackden Brook (Map 2A) between 120890 and 130880.
 (ii) The River Noe (Map 2A) between 133870 and 140850.
 (iii) The Stour Valley (Map 2B): Hambledon Hill 848123 (190 m) to the spot height at 812094 (225 m).

Remember that the scales of these maps are different.

2.22 What evidence can you gain from the map to show that the River Stour (Map 2B) is in its lower middle course, the River Noe (Map 2A) in its upper middle course, and the Grinds Brook (Map 2A) in its upper course? Study Fig. 2.1 to help with this exercise.

2.23 On Map 2B, Hambledon Hill (8412) and Hod Hill (8510) are outliers of the scarpland region. They have been separated from the main region lying east of the A350 by river erosion. Draw a sketch map to show these two hills, the scarpland, and the rivers Iwerne and Stour. Illustrate the relief using the following contours: 46 m, 61 m, 76 m and 152 m. Draw a line section along Northing 123 between the moat 821123 and Everley Hill Farm 889123. Carefully label the main physical features shown on this section.

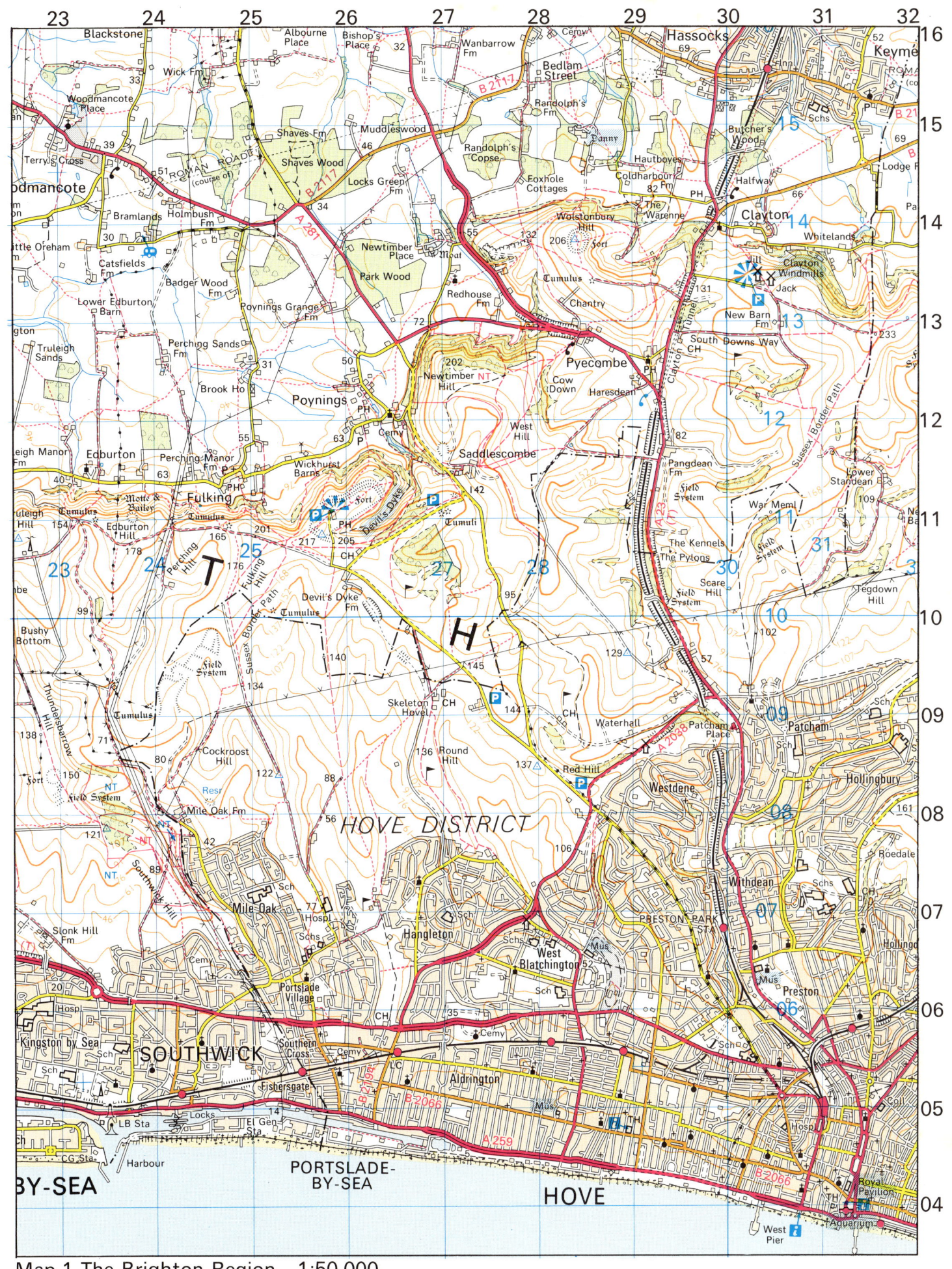

Map 1 The Brighton Region 1:50 000

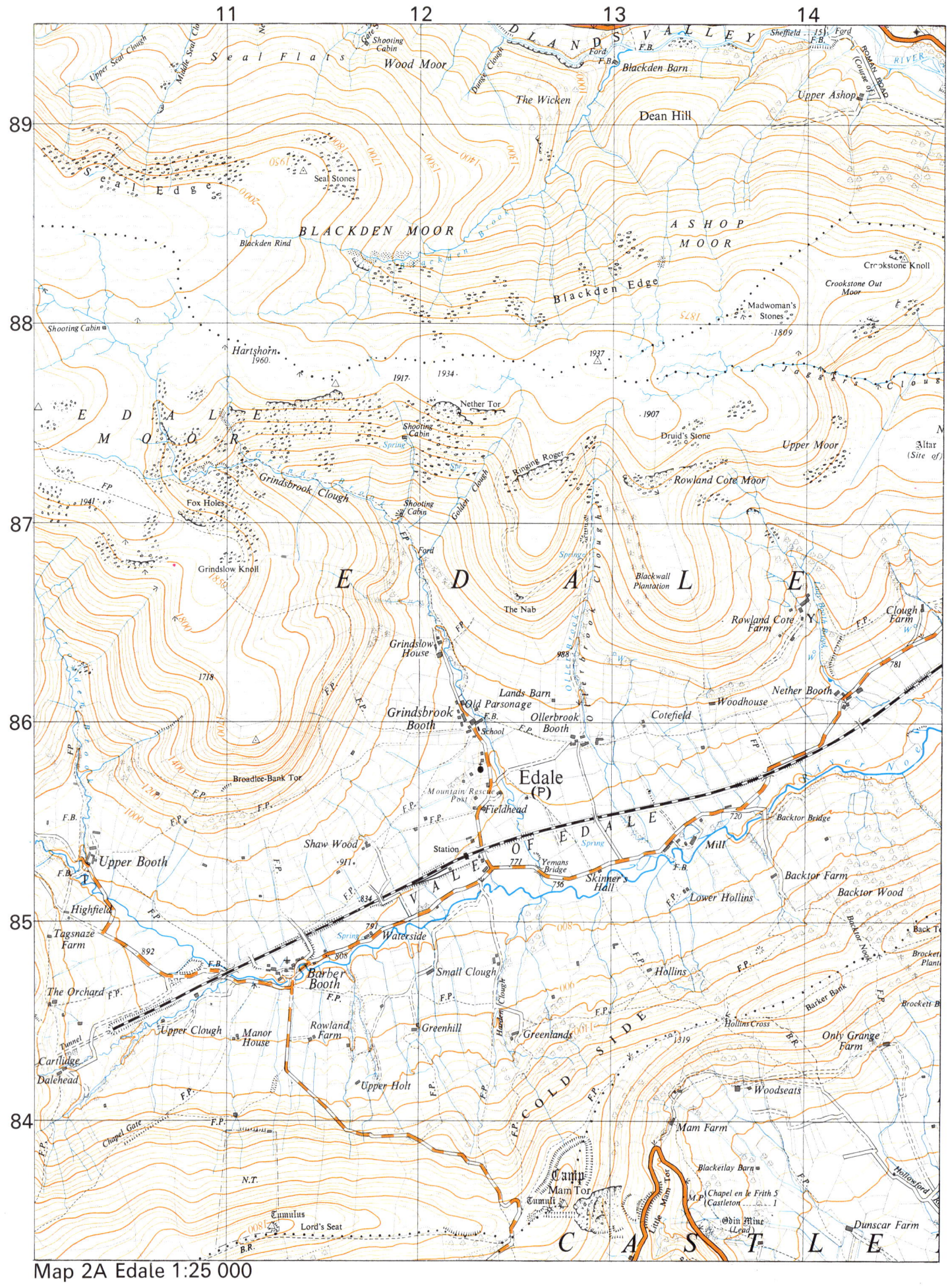

Map 2A Edale 1:25 000

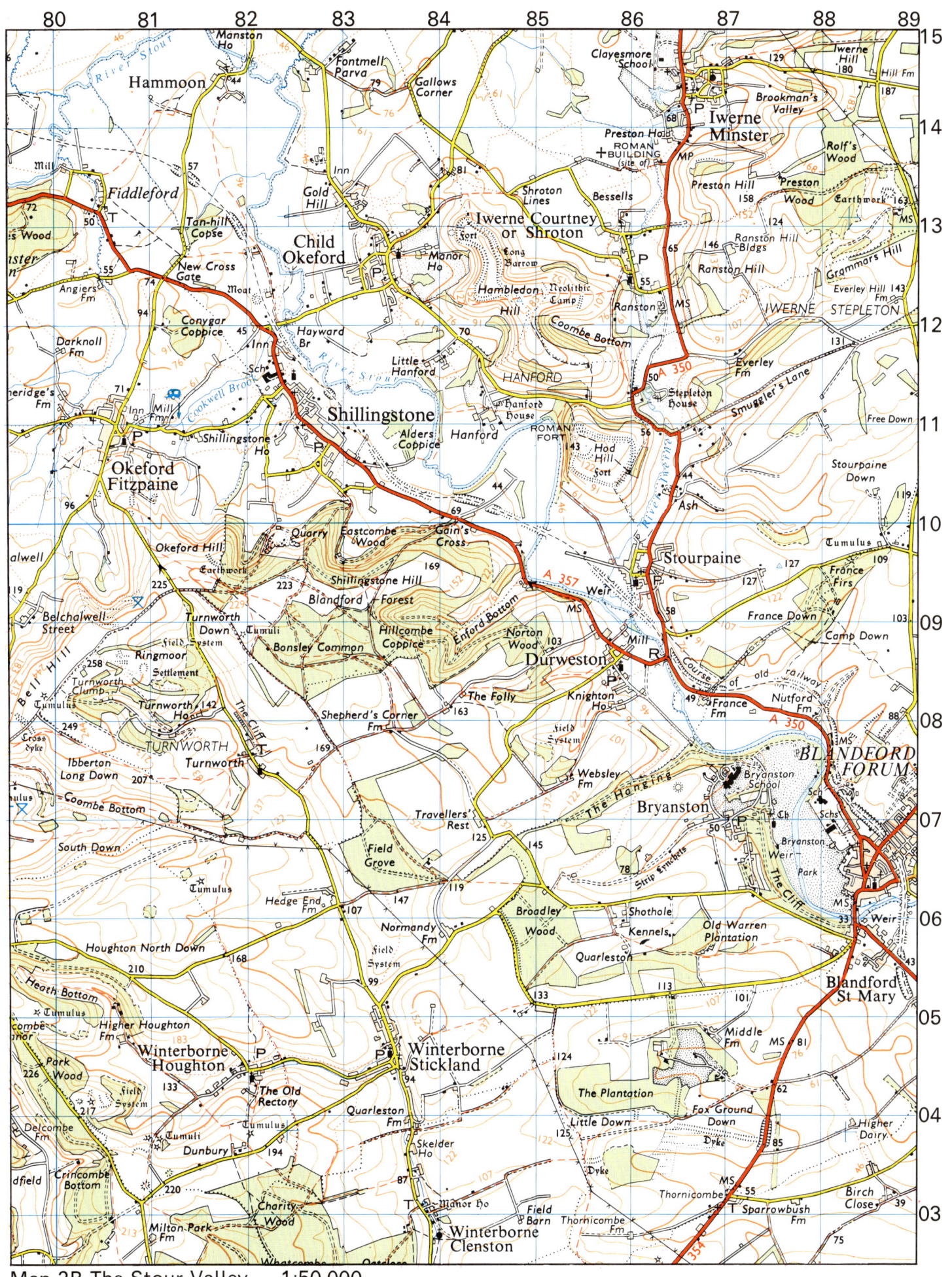

Map 2B The Stour Valley 1:50 000

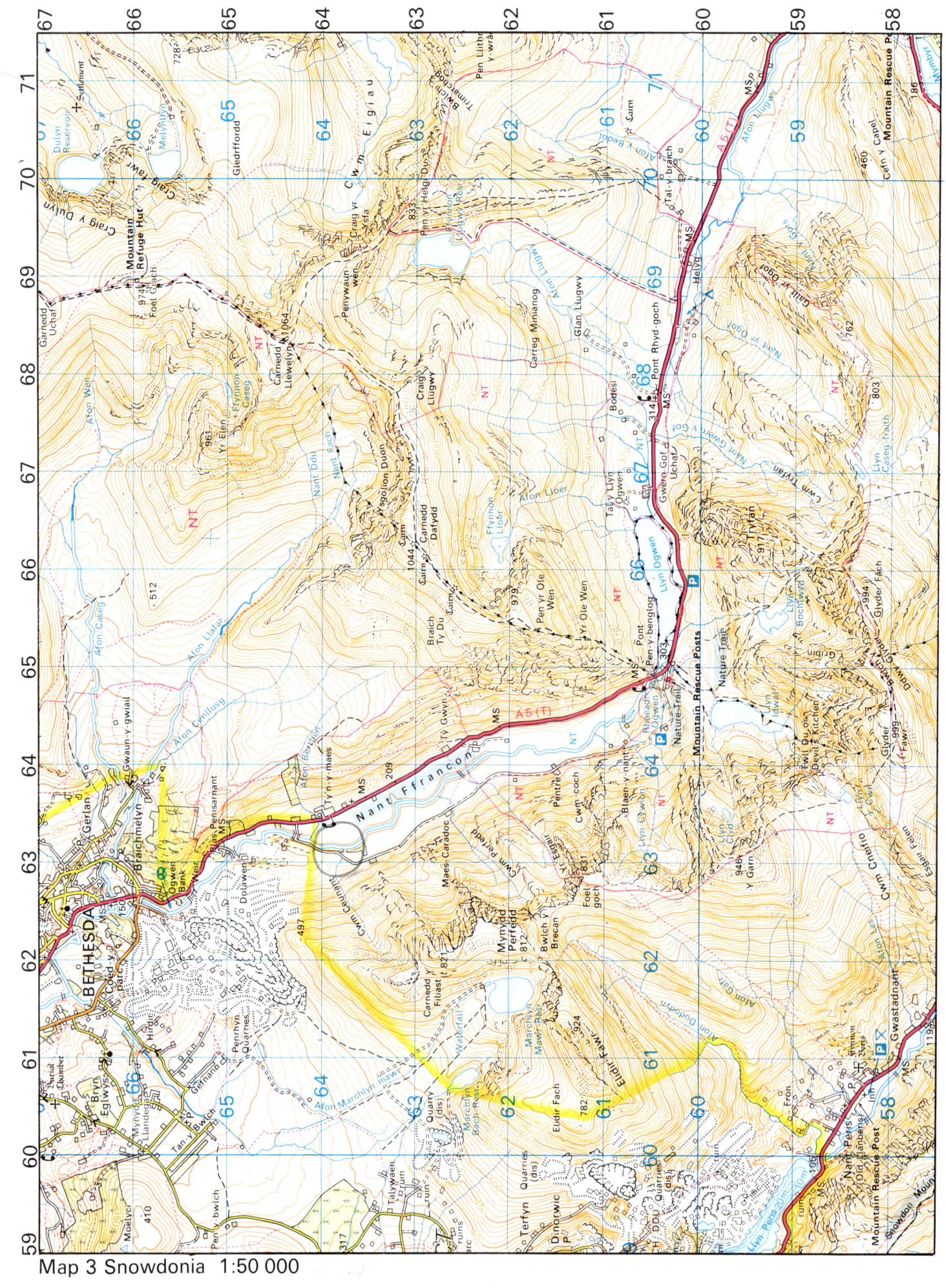

Map 3 Snowdonia 1:50 000

Plate 2B

2.24 Prepare a sketch map, at twice the scale of Map 2B, to show the following features in the vicinity of Bryanston School (8707):
  (i) the 46 m contour;
  (ii) the meander of the River Stour;
  (iii) the river cliff on the north-western edge of the spur on which the school stands;
  (iv) the more gentle slope on the south-eastern edge of the spur;
  (v) the floodplain.
  Use this diagram to illustrate the process by which a river erodes the floor of its valley, produces a flood plain and straightens the valley.

2.25 Name the hill marked A on Plate 2B, and answer these questions:
  (i) What is a hill fort?
  (ii) Who would have occupied this fort and when?
  (iii) Prepare a sketch map, with selected contours, to illustrate the natural features which would have helped in its defence. Indicate the defences added by the early occupiers.

2.26 Examine Map 2B and Plate 2B and identify three types of countryside in which woodland is found.

2.27 Describe the activities shown at B and C on Plate 2B. What is the significance of these activities to the geographer?

2.28 Bryanston School is situated at 8707. Examine Plate 2B carefully and suggest which of the following outdoor activities are available at the school: cricket, athletics, swimming, rowing, rugby, association football, horticulture.

2B.3 · Lower course

Study Map 7 (Lower Teesside) and Fig. 7.2, a line section drawn across it. The slopes on the northern side of the River Tees are very gentle and are typical of those found in the lower course of a river valley. South of the wide floodplain on the right bank, however, the slopes become steeper. These gradients probably relate to the occurrence of more resistant rocks.

Both the main River Tees and its 'old course' (4518, 4617 and 4618) and the creek in 5025 show the meandering characteristics of a mature river in a broad, flat, floodplain.

2C · The drainage system: guidelines for description

The following guidelines are given to help you describe and interpret the characteristics of a stream or river, the shape of its valley, and its relationship to other streams and rivers. When you have read these guidelines answer questions 2.29–2.31.

2C.1 · Course of a stream/river

1. In what general direction does it flow? What is the gradient of the stream? (If it proves difficult to read the contour numbers, then how frequently do contours cross the stream? What is their shape? e.g. V-shaped).

2. Measure the length of the stream. How does its width vary on the map? (Remember, however, that the width is not drawn true to scale, except in estuaries.)

3. Do the tributaries change the general direction of the stream? Imagine you are facing downstream. How many left and right-bank tributaries can you count? If there is a disproportionate number from one side rather than the other, can you explain why? How does the angle at which the tributaries enter the main stream change as you go downstream?

4. How does the form of the river change downstream? At what point does it start to meander? Do the features of the river suggest that it is changing from one stage in its course to a later one?

5. Describe any marked changes in the direction of flow. Can you explain these with regard to changes in relief or geology? Is there evidence of a river capture?

6. Are there waterfalls? Is there a reason?

7. Attempt an explanation for any lakes or artificial drainage.

2.29 What are the main differences between the courses of the River Wharfe (Map 4) and the River Tees (Map 7)

2.30 Compare the courses and valleys of the following:
(i) Park Gill Beck 9874 (Map 4) and Billingham Beck 4521 (Map 7);
(ii) Grinds Brook 1187 (Map 2A) and Blackden Brook 1288 (Map 2A).

2.31 Which of the following features do *not* appear in the valley of Blackden Brook 1288 (Map 2A):
(i) levees (ii) marshes (iii) ox-bow lakes
(iv) a confluence (v) interlocking spurs (vi) meanders?
Which of these features *do* appear in the valley of the River Noe (Map 2A)?

2C.2 · The nature of a valley

Since river valleys are rarely of constant width, depth or direction, we must rely on average measurements when we describe these characteristics. We must seek answers for the following questions.

1. How long is the valley? What is the average gradient of the long profile? Are there marked changes in direction? If so, can you explain why?

2. What changes are there in the width of the valley (see Fig 2.2)? Note any significant changes in the average gradient of the valley sides. Illustrate your answer with line sections. Relate your observations to changes in other features of the physical geography.

3. Are there abrupt changes in the level of the valley floor?

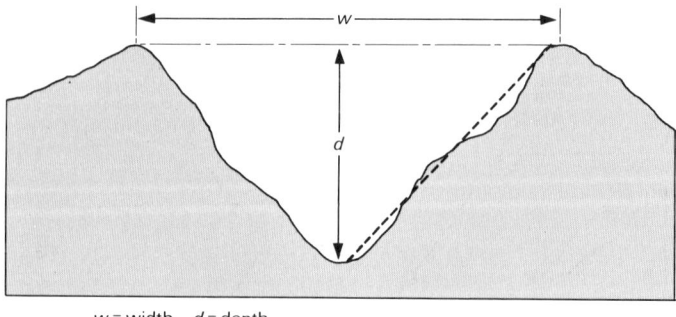

Fig. 2.2 The width and depth of a river valley

2.32 Describe the physical differences between the north and south sides of the valley of the River Noe on Map 2A. Can you explain the words 'Cold Side' in grid square 1284?

2.33 Give reasons to explain the route taken by the now disused railway line on Map 2B (Stour Valley) between 795140 and 890051.

2.34 Describe the valley of the Dowber Gill Beck (9872) above the confluence at 974725 (Map 4).

2.35 Use the guidelines in Chapter 9, 9C.1, to describe the site and situation of Edale (1285) on Map 2A.

2C.3 · Drainage networks

It is convenient to organise the description of a drainage basin in the following way.

1. Name all streams and state their direction of flow.
2. Consider the relative widths of the rivers and streams. Estimate their average gradients and attempt a comparison of the velocity and relative power of each one.
3. Examine the courses of the rivers and streams. Do they meander? Can you divide the rivers and streams into groups of those in their upper, middle, and lower courses? Have any meanders been transformed into ox-bow lakes?

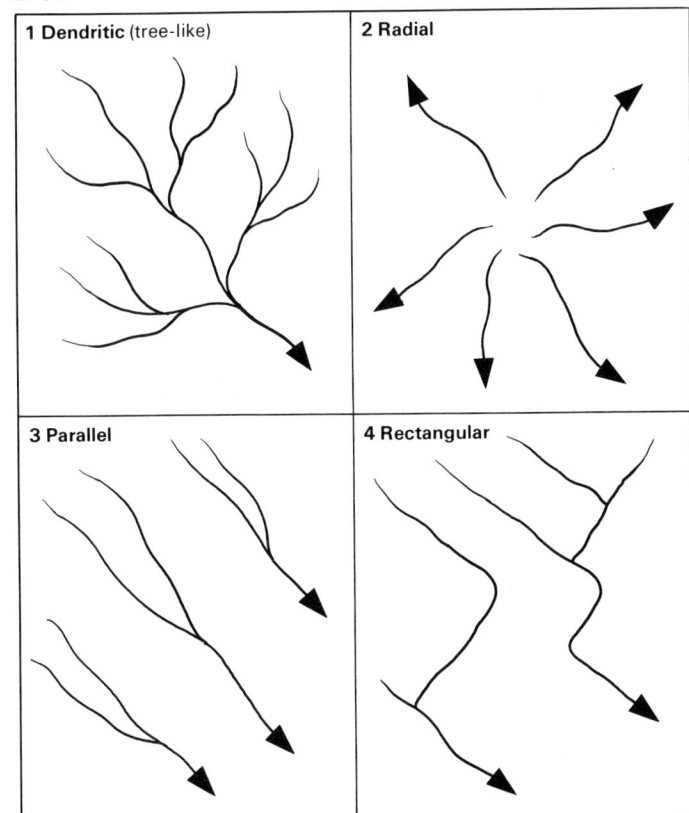

Fig. 2.3 Drainage patterns

4. Is there a distinctive geometrical pattern in the drainage system? Can such adjectives as (i) radial (ii) rectangular (iii) parallel (iv) dendritic (tree-like) be used to describe the pattern? Refer to the illustrations in Fig. 2.3 to help with this exercise.

5. Is there evidence of a river capture? If so, which river has been captured? Can you explain the capture in terms of topographical features, as demonstrated in Section 5A.3 (p.50)?

2.36 What differences in the pattern of drainage would you expect in the following environments:
 (i) a dome-shaped hill composed of impermeable, horizontally bedded rocks?
 (ii) a gently sloping region of similar rocks?
Draw sketch maps and diagrams to illustrate your answer. For guidance, refer to Fig. 2.3.

2.37 Refer to Map 5 (South Devon) and read Chapter 5, Section 5A.2. Describe the drainage system on the southern plateau, east of the Kingsbridge Estuary. Which of the illustrations in Fig. 2.3 best describes this pattern of drainage?

2.38 Select appropriate adjectives to describe the drainage systems in the areas listed below:
 (i) the dip-slope of the South Downs (Map 1 and Fig. 1.15);
 (ii) west of Nant Ffrancon and south of Northing 65 on Map 3 (Snowdonia);
 (iii) the area to the north and east of the A5 on Map 3 (Snowdonia).
To help with these answers, make a tracing of the drainage network in each area.

2.39 The word 'Winterborne' occurs in three places in the south-west part of Map 2B (Stour Valley). What does this word mean? Draw a simple sketch map of the valley system in which the villages are situated. Can you explain the origin of these valleys?

2D · Techniques for the comparison of river basins

2D.1 · Area of a drainage basin

A drainage basin is defined as that area of land which is drained by the trunk stream and its tributaries. Adjoining river basins are separated by a watershed. This is the ridge or area of land which divides the flow of surface water between the neighbouring basins.

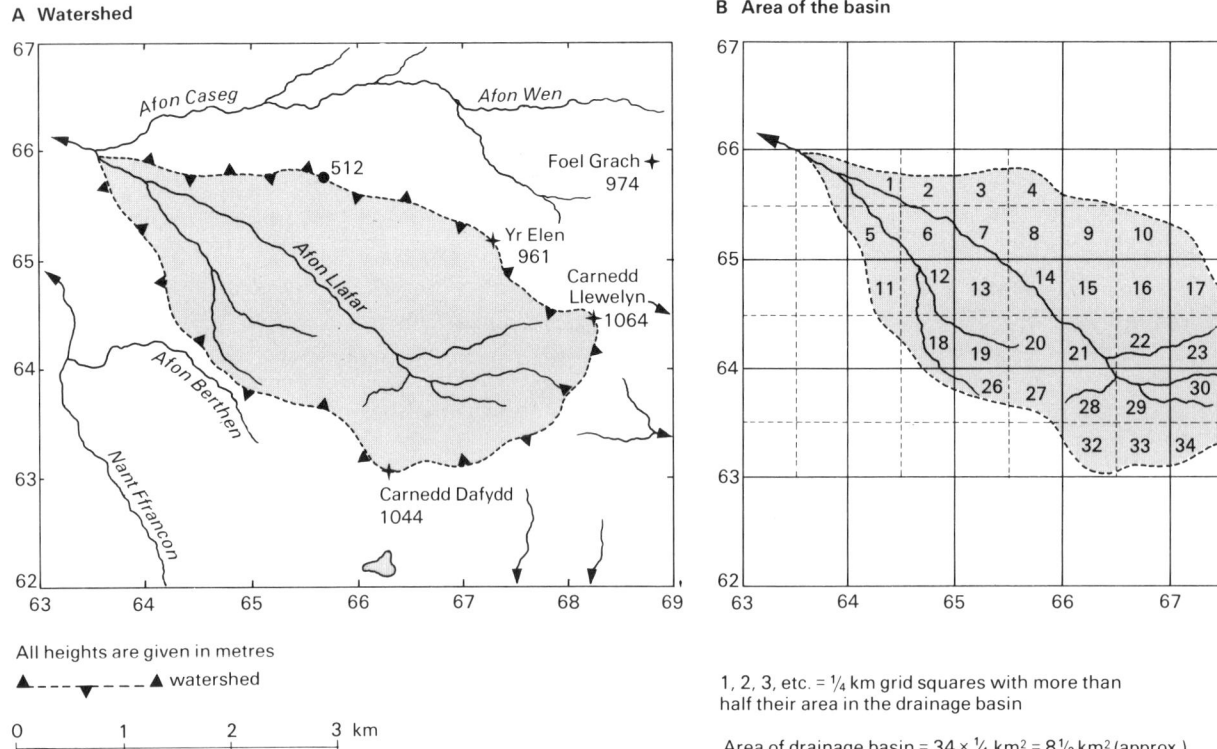

Fig. 2.4 The drainage basin of the Afon Llafar

Figure 2.4A shows the watershed of the Afon Llafar, upstream of its confluence at 636661 (Map 3, Snowdonia). Compare the position of this watershed with the detailed pattern of contours and spot heights shown on Map 3. Note the position of the highest points such as Yr Elen 673652 (961 m) and Carnedd Dafydd 662630 (1044 m). Remember, however, that the watershed may be an area of land rather than a clearly defined, steep-sided, ridge. Now refer to Map 2A (Edale) and follow the parish boundary along the high ridge through Hartshorn (111878), and the triangulation pillar 1937 at 129878. Notice how this ridge forms the watershed between north and south flowing streams along most its length. Parish boundaries coincide with watersheds, too, on Map 3 (Snowdonia), as for example between Carnedd Llewelyn 6864 and Carnedd Dafydd 6663.

The area of a drainage basin can be calculated by counting the number of grid squares it covers. In the case of a small drainage basin, however, like that of the Afon Llafar, a more accurate measurement can be gained by dividing the kilometre grid squares on the map into quarters. Then all those ¼ km squares which have at least half their area within the watershed are counted. As Fig. 2.4B shows, the area of this drainage basin is 8.5 km². Of course, with 1:25 000 maps like those for Edale (Map 2A) and Wharfedale (Map 4), measurements can be made for very small drainage basins by dividing each kilometre grid square into, say, twenty-five smaller squares, or even a hundred, if very precise calculations are needed.

2.40 Trace the drainage basin of the Blackden Brook, upstream from its confluence with the River Ashop at 132894, from Map 2A (Edale). Mark its watershed. Measure its area.

2D.2 · Drainage density

This value can be calculated using the formula below:

$$\text{Drainage density} = \frac{\text{total length of all streams}}{\text{total area of basin}} = \frac{L}{A}$$

where A = area of basin, L = total length of streams in that basin.

In this calculation it may be difficult to measure, accurately, the length of a stream. There can be a seasonal variation in the position of its headwaters arising from changes in the level of the water table. Therefore, the source is

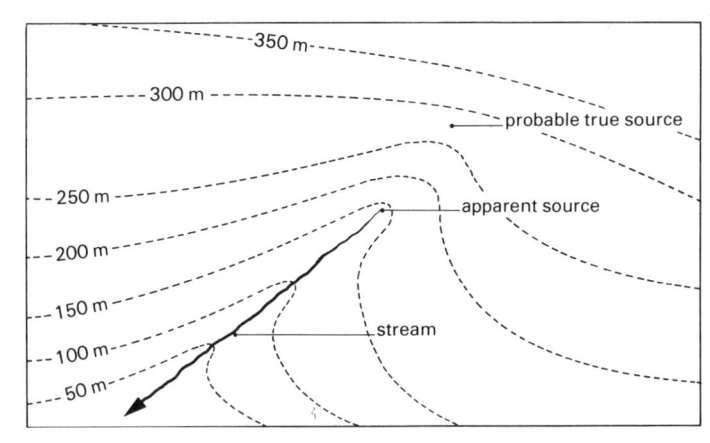

Fig. 2.5 Location of the true source of a stream

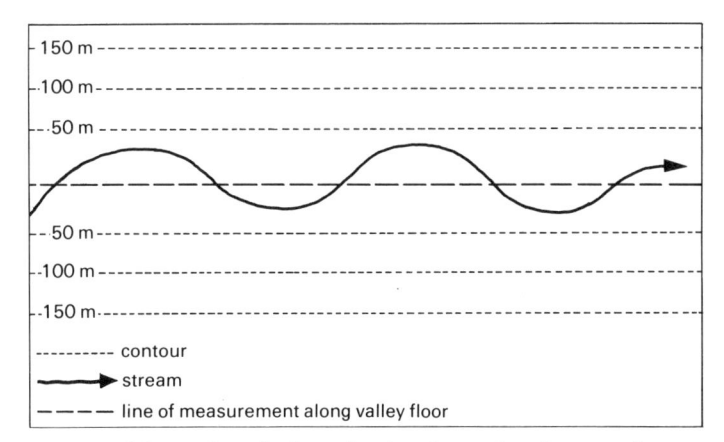

Fig. 2.6 Measuring the length of a river when it meanders within the lowest contour of the valley

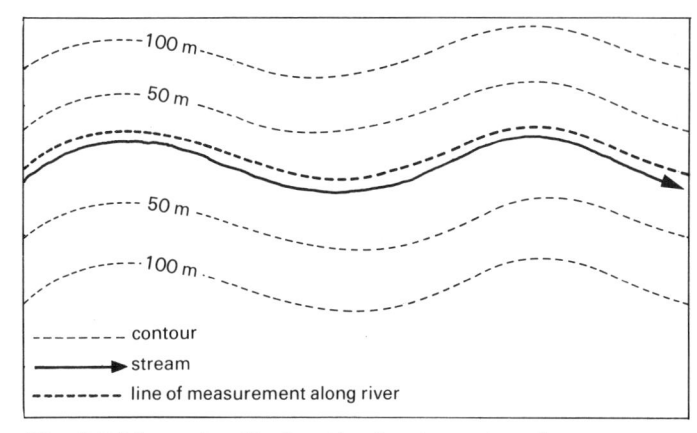

Fig. 2.7 Measuring the length of a river when the contour pattern follows the course of the river

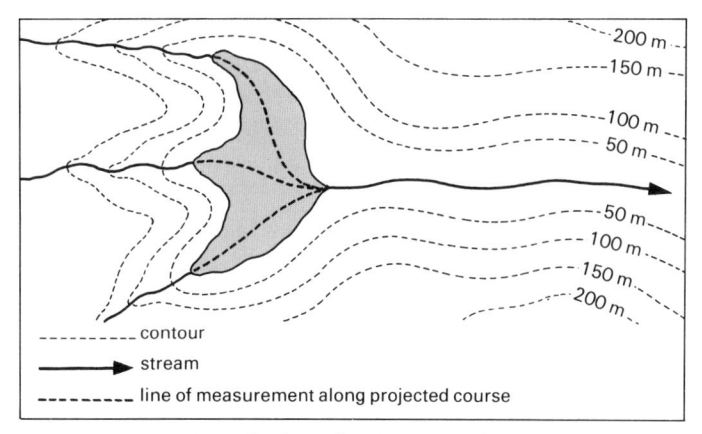

Fig. 2.8 Measuring the length of a river when it flows through a lake

normally taken as the highest contour which V's upstream, as shown in Fig. 2.5. In addition, other measurements are normally taken, as follows:

(i) As Fig. 2.6 shows, where a stream meanders within a valley, take measurements along the valley floor.

(ii) Where contours follow the winding course of the river, as in Fig. 2.7, take measurements along the river.

(iii) Where a river enters a lake, measure along the presumed direction of flow to the exit, as indicated on Fig. 2.8.

2.41 Compare the drainage basins of the Cam Gill Beck and Park Gill Beck above their confluence at 977740 on Map 4 (Wharfedale). For each basin calculate (i) its area (ii) its drainage density.

2D.3 · Average gradient

This is a useful calculation which suggests, in a general way, the velocity and power of the stream.

$$\text{Average gradient} = \frac{H - h}{L}$$

where H = height of source, h = height of outfall, L = length of main trunk stream.

Average gradients can be used as a basis for comparing different orders of stream in an area, or for comparing streams in one drainage basin with those in another.

2.42 Compare and contrast the courses, valleys, and drainage basins of the Grinds Brook (Map 2A, upstream of 127852) with the Park Gill Beck (Map 4, upstream of 978740). For each basin calculate (i) its area, (ii) its drainage density, (iii) the average gradient of the stream.

2.43 Trace the drainage basin of the Afon Llafar above the confluence at 636661 on Map 3 (Snowdonia). Calculate (i) the average gradient of the Afon Llafar (ii) the drainage density of the basin. Refer to Fig. 2.4B for the area of the drainage basin.

2.44 Which valley would provide the better site for a reservoir, the Nant Ffrancon (Map 3) or Wharfedale (Map 4)? Support your answer with appropriate evidence from the map. What factors would water board engineers need to consider before building such a reservoir?

CHAPTER 3
Snowdonia

Chapter Plan

3A · **Location of a map extract**

3B · **Landforms in a glaciated upland region**
    3B.1 Characteristic landforms
    3B.2 Physical regions

3C · **Human activity in an upland environment**
    3C.1 Economic activity
    3C.2 Recreation and the physical environment
    3C.3 Settlement patterns

3A · Location of a map extract

With but few exceptions, the system used in Britain for numbering roads is that indicated on Fig. 3.1. All 'A' and 'B' class roads are numbered so that those situated between the A1 and A2 will start with 1, those between the A2 and A3 with 2, and so on, in a clockwise direction. The map reader can use this information to identify that part of the country shown on the Ordnance Survey map.

Language can also provide a clue to location. This chapter uses a map extract lying to the north east of Snowdon (Snowdon Mountain Railway 590575) in North Wales. The A5 crosses the area diagonally. Examine the list of Welsh words (and English translations) in Table 3.1 as a guide to interpreting the physical landscape.

Table 3.1 Landscape features: English translations of selected Welsh words

Afon River	*Fawr* big
Bwlch pass	*Ffynon* well, spring
Blaen end	*Llyn* lake
Coed wood	*Moel* hill
Cwm valley	*Nant* stream
Carnedd ⎫ mountain	*Pen* head, summit
Mynydd ⎭	*Pont* bridge
Du black	*Rhaedr* waterfall
Fach small	

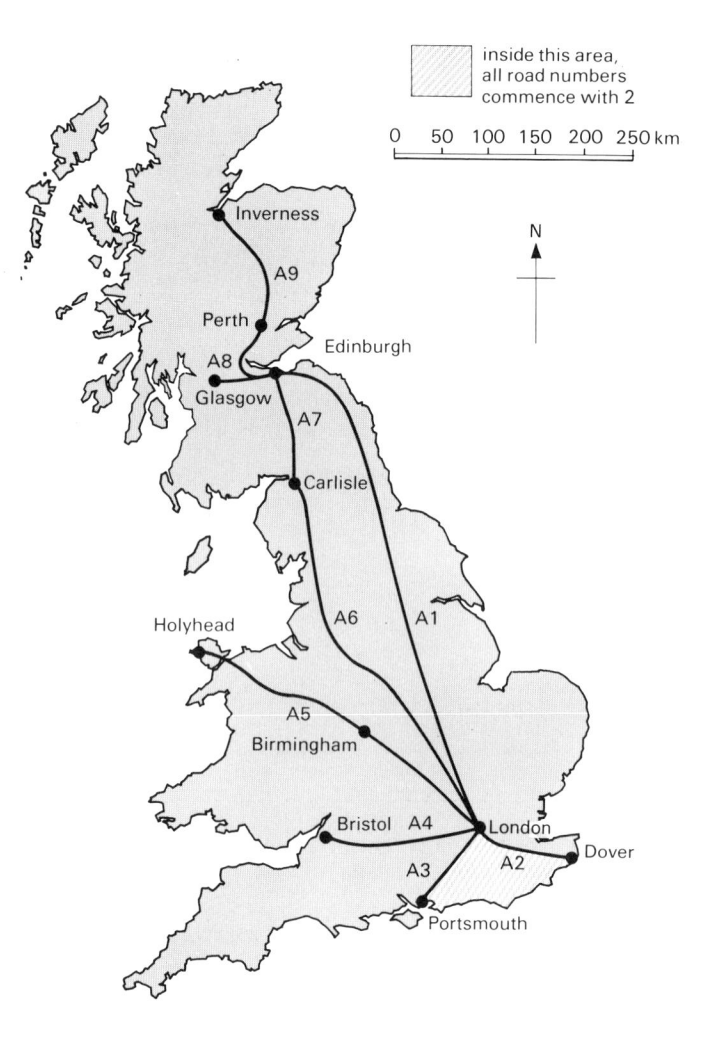

Fig. 3.1 Road numbers and map location

3B · Landforms in a glaciated upland region

This study based on Map 3 is concerned with two themes. Firstly, it examines a typical range of landforms found in a glaciated upland region of Britain. Secondly, it interprets the related pattern of settlement and land use.

3B.1 · Characteristic landforms

Corries. During the Ice Ages the main glacier which occupied Nant Ffrancon was supplied with the ice from several corries, including those occupied by Llyn Idwal (6459), Llyn Bochlwyd (6559) and Ffynon Lloer (6662). Although in detail the shapes of these corries are different, as they have been modified since the Ice Age, they have several features in common. They are

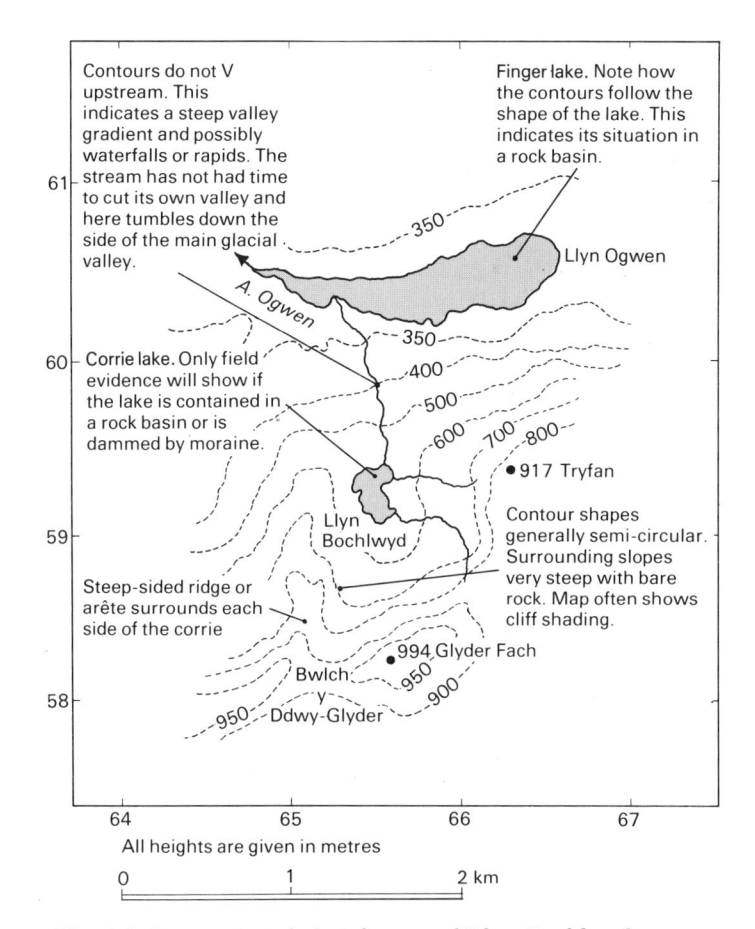

Fig. 3.2 An annotated sketch map of Llyn Bochlwyd

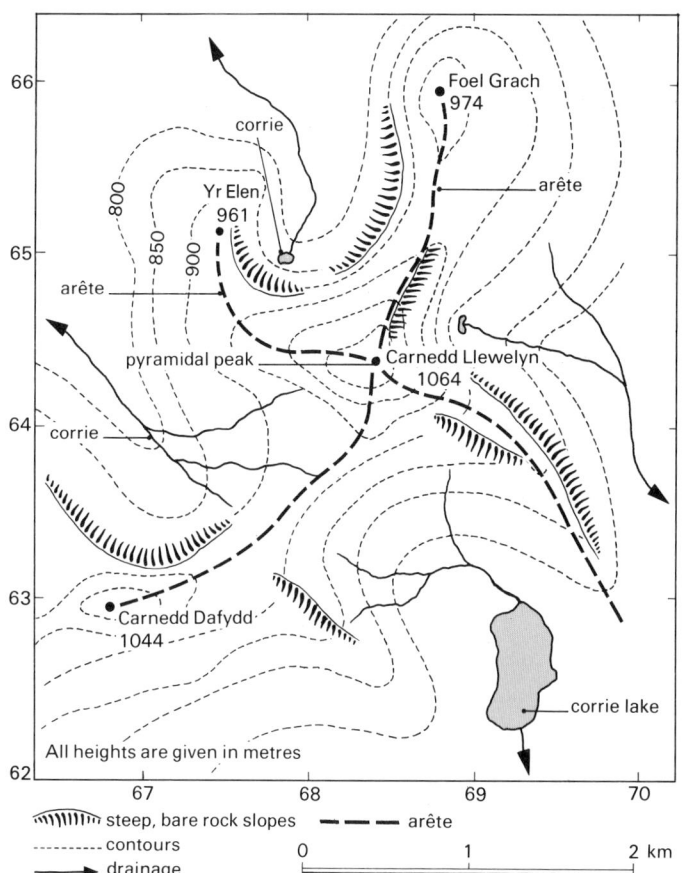

Fig. 3.3 Carnedd Llewelyn: a sketch map to illustrate a pyramidal peak

generally semi-circular in form, surrounded by high walls with steep slopes and are separated from neighbouring corries by narrow, steep-sided ridges or arêtes. Figure 3.2 illustrates these characteristics from Llyn Bochlwyd (6559).

Not all corries now contain lakes, however. Cwm Tryfan (6658) is an example. It is semi-circular in shape, with steep bare-rock walls, as is Llyn Bochlwyd (6559). Several less well-developed corries on the western side of Nant Ffrancon, for example Cwm Ceunant (6263), Cwm Perfedd (6362) and Cwm Coch (6361), also supplied ice to the main valley glacier.

Pyramidal peak and arête. Study Fig. 3.3. This sketch map shows corries surrounding a pyramidal peak, Carnedd Llewelyn (6864). What processes during the Ice Ages produced these glacial landforms? Imagine a prolonged period of sub-zero temperatures with ice eroding the headwall and sides of the corries. Later, in a relatively less severe climate an alternating sequence of freezing and thawing of the rock would take place. In conjunction with mechanical weathering, these processes would steepen the ridges or arêtes which separate neighbouring corries. Figures 3.4A and 3.4B show that these processes have led to the progressive retreat of the corrie head-walls and sides. Carnedd Llewelyn has thus emerged as a pyramidal peak.

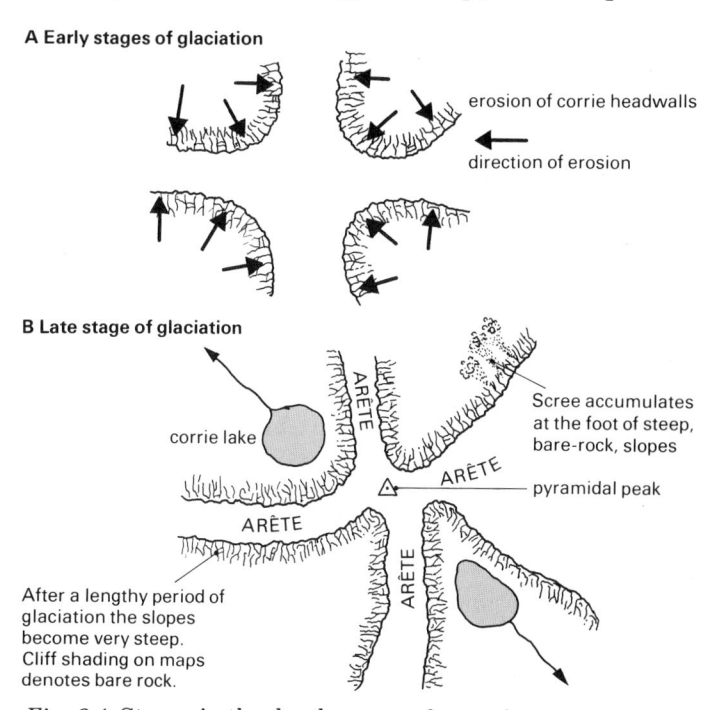

Fig. 3.4 Stages in the development of a corrie

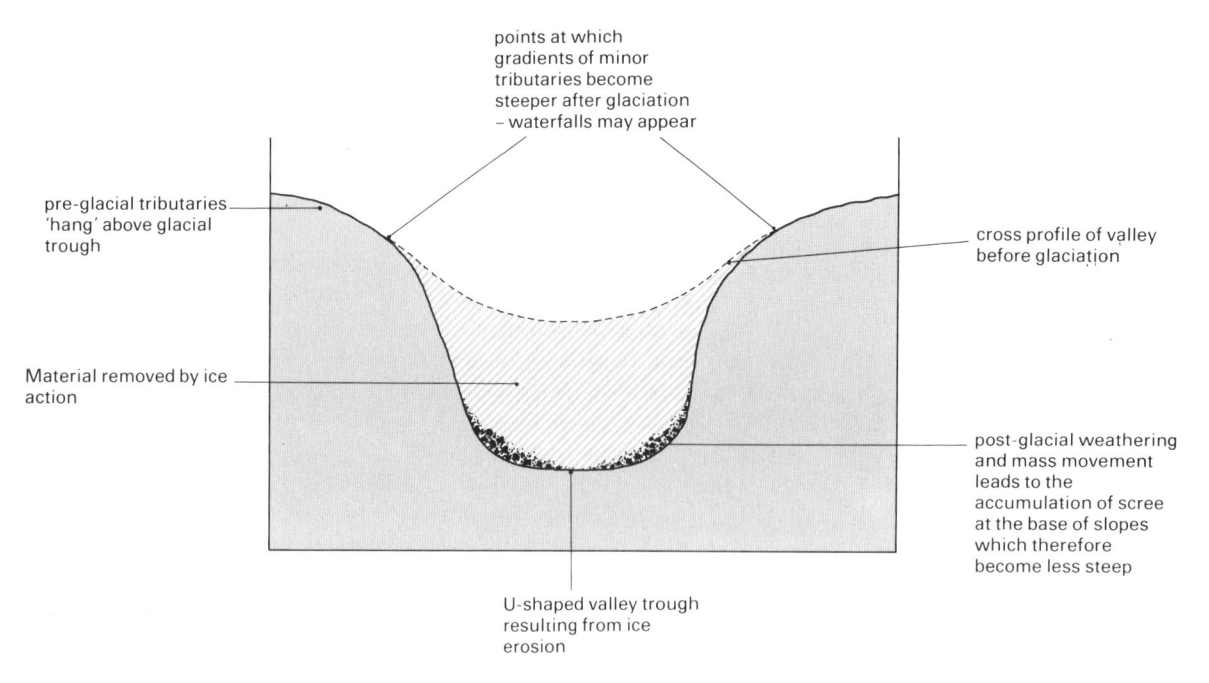

Fig. 3.5 An annotated diagram of a typical U-shaped glacial valley

The map shows a pattern of closely spaced contours which indicates steep slopes, with bare rock surfaces and cliff faces encircling the corries (Fig. 3.3). There is also evidence of scree slopes. These confirm the mechanical weathering of steep rock faces in the post glacial period (6765; 6963; 6865).

Glacial U-shaped valley. Nant Ffrancon is a classic example of a glaciated valley (Fig. 3.5). It has a flat floor and very steep sides. Between 632647 and 646607, a distance of 4.1 km, only two contours cross the valley floor. Assuming a fall of 20 m, this gives a longitudinal gradient of 0.5% ($\frac{20}{4100} = \frac{1}{205} = 0.5\%$ approximately). Rhaedr Ogwen (647605) marks the prominent end wall of the valley. Here, as Fig. 3.6 shows, the erosive power of five small glaciers was combined to deepen the lower valley and straighten its sides.

Hanging valley. Try to reconstruct the environment of the Nant Ffrancon region during the Ice Ages, by answering the following questions.

(i) Which way do the slopes on either side of the valley face?
(ii) Which side of the valley would have been the more sheltered from the direct rays of the sun during the period of corrie glaciation?
(iii) On which slopes would the snow and ice have remained for the longest period?
(iv) Now explain why corries are found only on one side of Nant Ffrancon.

Figure 3.7 summarises the distinctive characteristics of a hanging valley. Such valleys are the result of a process in which the main valley becomes over-deepened by ice action and tributary valleys are left at a higher level. What evidence can you find on the map to confirm this process?

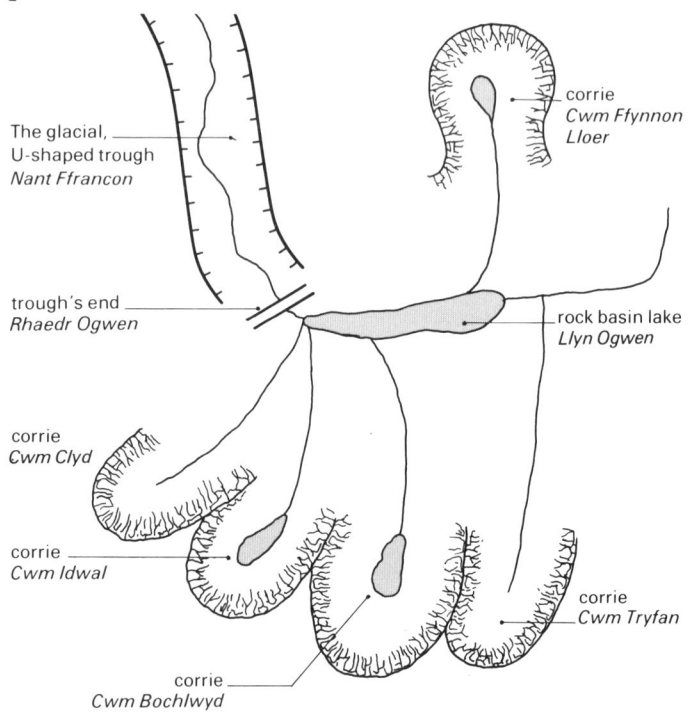

Fig. 3.6 Sketch map of Nant Ffrancon to show glacial features

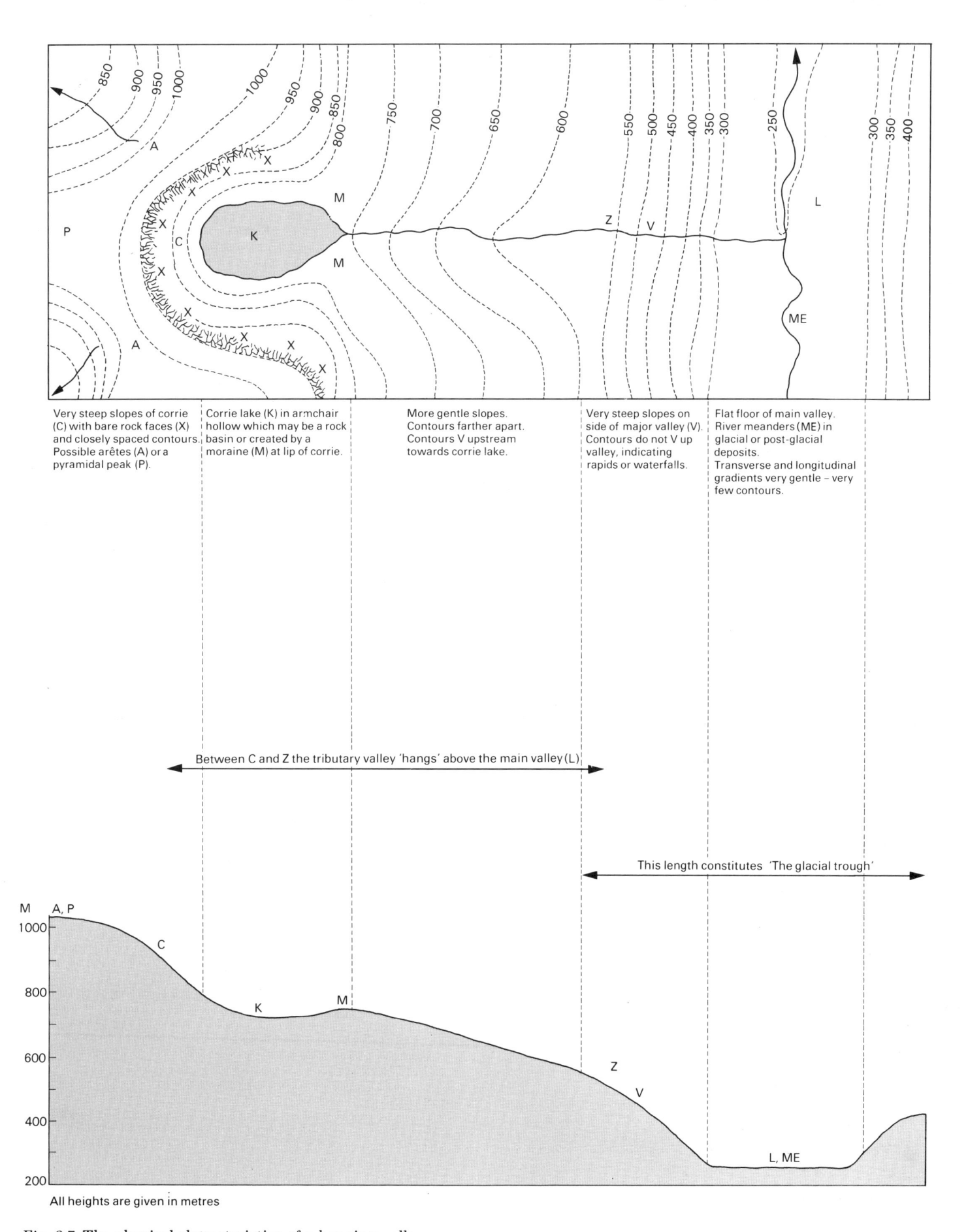

Very steep slopes of corrie (C) with bare rock faces (X) and closely spaced contours. Possible arêtes (A) or a pyramidal peak (P).	Corrie lake (K) in armchair hollow which may be a rock basin or created by a moraine (M) at lip of corrie.	More gentle slopes. Contours farther apart. Contours V upstream towards corrie lake.	Very steep slopes on side of major valley (V). Contours do not V up valley, indicating rapids or waterfalls.	Flat floor of main valley. River meanders (ME) in glacial or post-glacial deposits. Transverse and longitudinal gradients very gentle – very few contours.

Between C and Z the tributary valley 'hangs' above the main valley (L)

This length constitutes 'The glacial trough'

All heights are given in metres

Fig. 3.7 The physical characteristics of a hanging valley

3.1 Examine Plate 3 (see Fig. 3.10 for location).
(i) Identify a hanging valley.
(ii) Locate Blaen-y-nant (6360) and Maes Caradoc (6362).
(iii) Identify the peaks A, B; valley X; landform C; land uses D, E, F; hill G.

3.2 Refer to grid square 6361. Examine the detail shown for this area on Plate 3, and answer the following questions.
(i) In what respects is the air photograph superior to the map in showing detail of the physical environment?
(ii) What are the main limitations of the air photograph as an information source for the physical geographer?

3.3 With regard to the cross profile of the Ogwen valley, explain the significance of the line marked Y–Y on Plate 3. (For a clue, examine Fig. 3.5 and 3.7)

3.4 Compare the appearance of the surface terrain on Plate 3 with that shown on Plate 1A. What major contrasts do you observe? How do these relate to major differences in rock type?

3.5 Describe the land-use pattern on the floor of Nant Ffrancon. How does the map complement the detail shown on the air photograph?

3.6 Comment on the alignment of the district boundary between Garnedd Uchaf (6866) and Carnedd Llewelyn (6864). What physical feature does this boundary follow?

3B.2 · Physical regions

A set of guidelines and techniques for identifying physical regions in the landscape are examined in Chapter 1, Section 1B. This study focuses on the

Plate 3

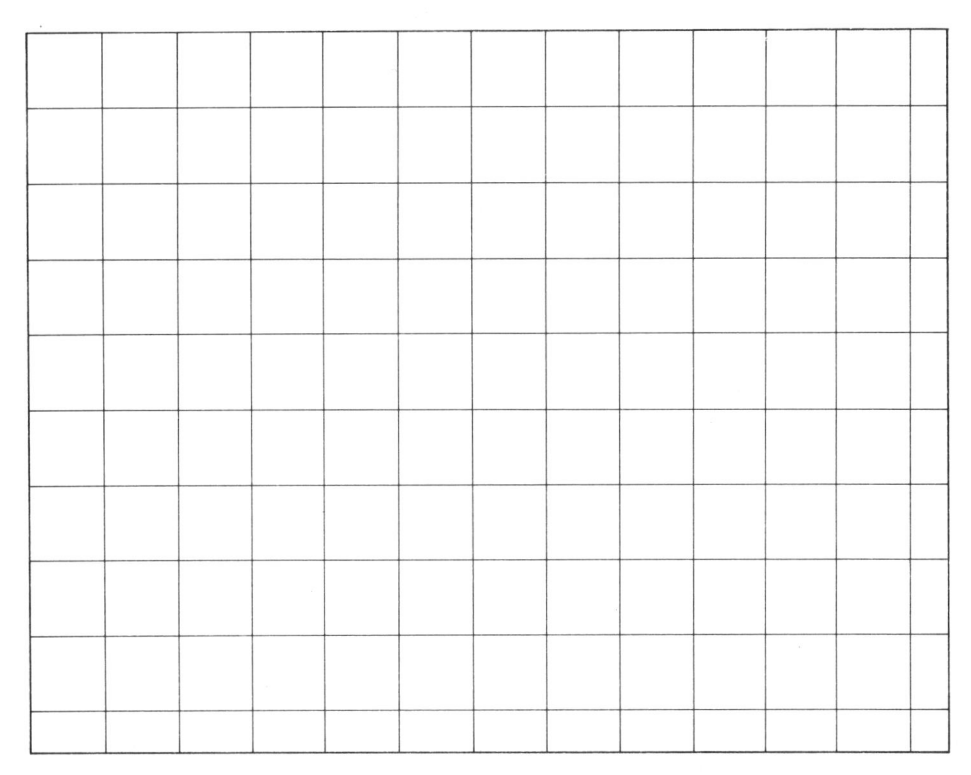

Fig. 3.8 Graticule for drawing a sketch map

role of the sketch map as a means of organising and presenting significant features in the landscape.

Preparation of a sketch map. Draw a graticule or grid, like Fig. 3.8, which covers the area of Map 3. Mark all values for northings and eastings. Answer the following questions, and use your answers to locate the major landforms on the graticule.

3.7 What is the total area of the map in km²?

3.8 Mark the position of the 350 m contour. What proportion of the area lies above 350 m?

3.9 Find the main watersheds.

3.10 Find the highest point and note its height. Locate other points above 800 m.

3.11 How many glacial valleys dissect the mountains?

3.12 Identify, and name, examples of corries, arêtes, pyramidal peaks, U-shaped valleys and hanging valleys.

Description and interpretation of relief regions. Figure 3.9 identifies three major river basins: the Afon Ogwen drains two-thirds of the area to the north west; the Afon Llugwy drains the eastern part of the region; while in the extreme south west the drainage flows to Llyn Peris (5959). The main watershed follows a line of high points from Foel Grach (689659) to Craig Llugwy (680630), crosses the valley between the Ogwen and Llugwy streams at 677604, to Glyder Fach (657583), Glyder Fawr (643580), Y Garn (630596), Foel Goch (628613) and terminates at Carnedd y Filiast (620628). Most of these points lie above 950 metres.

There are glaciated landforms throughout this upland region, namely corries, hanging valleys and pyramidal peaks. Furthermore, the surface gradients are characteristically steep and exposures of bare rock, cliff faces and scree slopes are common. This kind of evidence is widespread throughout the map area. It is therefore tempting to refer to the whole map as one region of glaciated upland. On a closer inspection of Fig. 3.9, however, it appears that several of these physical features are less well-developed or absent in the north west, especially below the 350 metre contour. Can this area therefore be legitimately defined as a separate region? Do the following exercises and summarise the evidence.

3.13 Examine the distribution of land below the 350 m contour on Map 3.

3.14 Draw line sections, and mark and annotate the major breaks of slope between the following points:
 (i) Mynydd Perfedd (623618), 812 m and 594659, 410 m.
 (ii) Carnedd Dafydd (662630), 1044 m and 625663, 150 m.
 (iii) Mynydd Perfydd (623618), 812 m and 656619, 979 m.

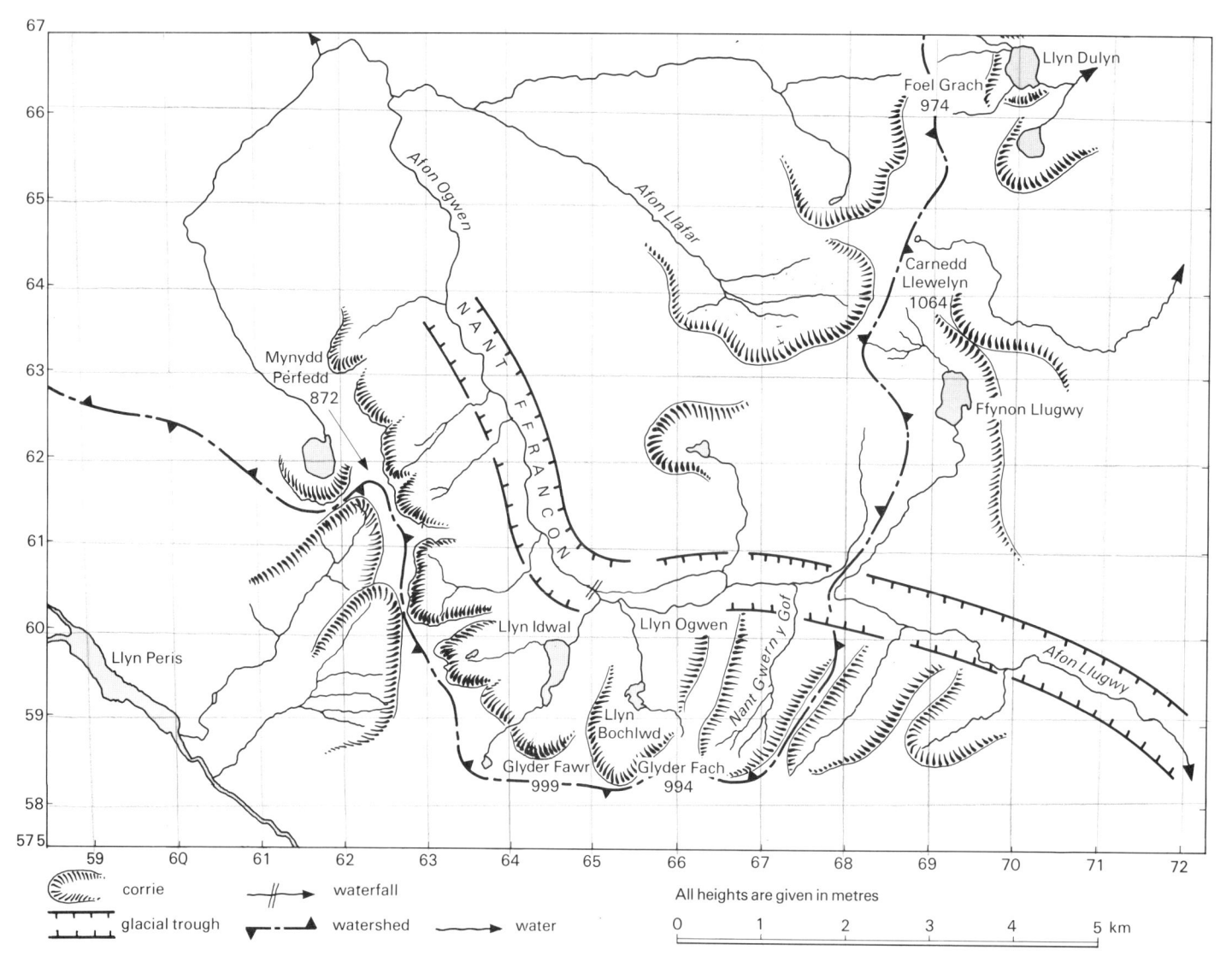

Fig. 3.9 Sketch map of glacial landforms

3.15 For the regions above and below the 350 m contour
 (i) Contrast the network of surface drainage;
 (ii) Examine the 'roughness' of the terrain.

These exercises show that there is a major break of slope between the north-west sector and the main core of the mountain region. It is difficult, however, to mark a precise boundary between these regions on the map. In addition, a narrow tongue of lower land extends along the Nant Ffrancon to the waterfalls at Rhaedr Ogwen, crosses the watershed and continues south east along the Llugwy valley. Throughout the north-west region, however, the wider spacing of contours indicates that slopes are less steep than elsewhere on the map. The surface terrain, too, is less rugged. Unlike the mountainous area, there are no lakes in the north west and fewer minor and steeply graded streams. These contrasts are therefore used as a basis for defining the north-west sector (Fig. 3.10) as a separate region in the physical landscape. Its general limits extend, approximately, to the line of the 350 m contour, and cut across the north-west end of Nant Ffrancon at Ty'n-y-maes (635640). Within the mountain core, the lower lying corridor of Nant Ffrancon and the Llugwy valley can be considered as a sub-region of the lowland area.

3C · Human activity in an upland environment

3C.1 · Economic activity

Agriculture. The general guidelines on agricultural land use in Chapter 1 (1D.3) are used to provide a basic statement on farming

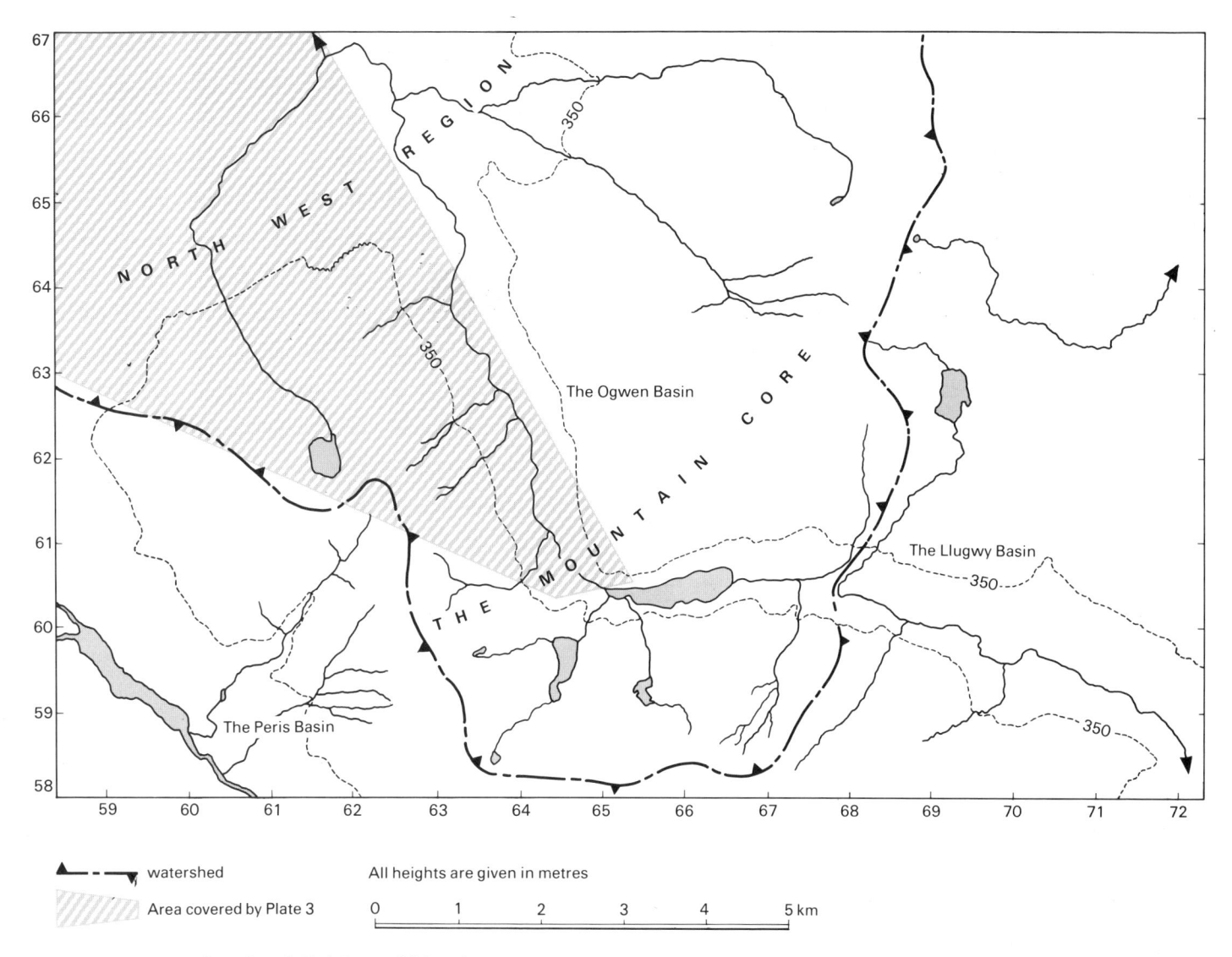

Fig. 3.10 Proposed regional divisions of Map 3

activities in Snowdonia. As Fig. 3.10 shows, the distribution of land below the 350 m contour is confined to the lower valleys of the Llugwy and Ogwen, and around Llyn Peris. Elsewhere the land surface is very steep, with most gradients exceeding 14%. There is also abundant evidence of bare rock. Thin soils probably occur on many slopes. In addition, the high rainfall in this area results in rapid runoff from steep slopes. This might possibly create a flooding problem on the flat floor of Nant Ffrancon, leading to the waterlogging of the soil in winter.

As a result, the agricultural economy is probably based on the raising of sheep and beef cattle on rough grass pastures with some feeding stuffs being imported into the area. In winter some livestock are probably moved to pastures at lower altitudes outside this area.

Forestry. Although the area is not heavily afforested, there are three major stands of private (not Forestry Commission) woodland in the north west. Each exceeds ¼ km^2, and lies below the 400 m contour. At Tan-y-bwlch (5965; 6065) it seems likely that trees have been planted as shelter belts to protect livestock from the north-east and south-west winds, and associated blizzards in winter. Only in grid square 5966 is there evidence of an access road within the woodland. This may serve the homestead of a farm which has been partially afforested.

Quarrying. Quarrying is a marked feature of the landscape in the south west (5959, 5960, 5961). In this area, it would appear that the industry has declined, with evidence of ruins and disused quarries. In contrast, in the north west the quarrying industry, which covers approximately 4 km^2 to the south west of Bethesda, appears to be active.

3C.2 · Recreation and the physical environment

Each year since 1945 an increasing number of tourists have made holiday visits to the mountain regions of Britain. Among the reasons for this trend are the rising levels of car ownership, longer holidays and the services provided by tour operators. Mountain environments offer a range of recreational opportunities: mountaineering, rock climbing, fell walking and winter sports appeal to many active groups, while others enjoy pony trekking, sailing and fishing. Many visitors, too, derive their pleasure by viewing mountain scenery from a lakeside car park or lay-by on a tourist route.

In some mountain areas the government and county authorities have prepared management plans to protect the landscape from the harmful effects of tourism. They have also provided information services and facilities to meet the visitors' basic needs. Snowdonia is such an area. Examine the map and answer the following questions.

3.16 Study the route taken by the A5(T) between 715593 and 619670.
  (i) Explain why the road follows this route.
  (ii) Locate the highest point on this route and note its height.
  (iii) How many times does the A5(T) cross the Afon Ogwen and the Afon Llugwy? Can you explain why this happens?

3.17 What is a mountain refuge hut (6965)? What is the purpose of a mountain rescue post (6460)?

3.18 What is a nature trail (6460)?

3.19 How many camping and caravan sites can you find on the map? Should visitors be free to camp wherever they would like? Which other locations would you recommend for new caravan sites. Give map references and explain why you have chosen these sites.

3.20 Find the broad yellow band between 632670 and 595590. What does this represent and what does it enclose? Why would those concerned with planning the countryside wish to designate such a special area? Why do you think Bethesda was excluded from this area?

3.21 Would you like to spend a holiday in this part of the country? Giving map evidence, name some of the activities you could pursue. What restrictions would you impose on the activities of visitors so that everyone can enjoy the amenities offered?

3.22 Hang-gliding has become a popular sport in Britain. Suggest possible sites for this activity, bearing in mind the requirement for road access and the nature of the sport.

3.23 Does the physical environment shown on the map lend itself to the establishment of a winter sports centre? Give reasons to support your answer.

3.24 You are the leader of a walking party. Your minibus has been parked in the car park in grid square 6058 and you are going to walk to the Nature Trail in grid square 6460. Describe the route you will follow. Give compass directions and the altitudes and gradients to be negotiated.

Your visit takes place late in September and the weather forecast is not good. What instructions would you issue for clothing, and who would you notify, before setting out?

3C.3 · Settlement patterns

Answer questions 3.25 to 3.29 and relate the basic distributions of settlement and communications to the physical geography of the region. First of all, you may wish to refer to Chapter 5, Sections 5C.1 and 5C.3, and Chapter 9, Section 9C.1 for guidance.

3.25 Are these statements true or false?
  (i) There is no settlement above 400 m.
  (ii) The altitude of the A5(T) road never rises above 314 m.
  (iii) Villages or hamlets are not found above the 250 m contour.

3.26 Trace the settlement distribution in the Llugwy and Ogwen valleys to the east of Easting 63. What sort of settlement is there? Where is it? What is the maximum height at which settlement is found? Why have the houses been built in these situations?

3.27 Examine Map 3 and the air photograph (Plate 3) and describe the pattern of settlement and communications in the Ogwen Valley between Rhaedr Ogwen (6460) and Ogwen Bank (6265). What is the relationship between the communication pattern and farming activity?

3.28 Bethesda has been described as an industrial village whose economy is based on slate quarrying (Penrhyn Quarries 6164).
  (i) Describe the differences between the housing at Tan-y-Bwlch and Douglas Hill in 6064.
  (ii) Suggest other occupations in which the working population of Bethesda may be employed.
  (iii) What is slate? For what is it used?

3.29 Give examples to show the effect of the tourist industry on the pattern of settlement.

CHAPTER 4
Wharfedale

Chapter Plan

4A · **Physical landscape**
    4A.1 Relative relief diagram
    4A.2 Geology and relief
    4A.3 Limestone landscape
    4A.4 Transect diagram

4B · **Human activity in the upland environment**
    4B.1 Economic activity
    4B.2 Settlement patterns and society

4A · Physical landscape

This study uses three techniques to interpret the physical landscape around Kettlewell (Map 4). First, a relative relief diagram is constructed to complement the pattern of contours on the map and to summarise variation in the height of the land surface. Secondly, information is introduced from a geological map to analyse the pattern of relief and to explain features of the landscape. Finally, a transect diagram is prepared to illustrate the relationship between the physical setting and distributions of vegetation, settlement and communications.

4A.1 · Relative relief diagram

Sometimes when interpreting the land surface of a region it is important to be able to identify local differences in relief. These variations in the height of the land can be summarised in a relative relief diagram like Fig. 4.1. This diagram shows the relative relief of the land surface in the vicinity of Kettlewell, and was prepared in the following way.

1. A grid of ¼ km squares, drawn on tracing paper, was superimposed on the map between Northings 715 and 725.

2. The number of contour lines crossing each square was counted, and noted on the tracing overlay. In some squares, for example 9771 and 9871, rock drawings have replaced contours and it was necessary to include values for the missing contours.

3. Figure 4.2 was prepared to show the distribution of contour crossings for the seventy-six squares in the area studied.

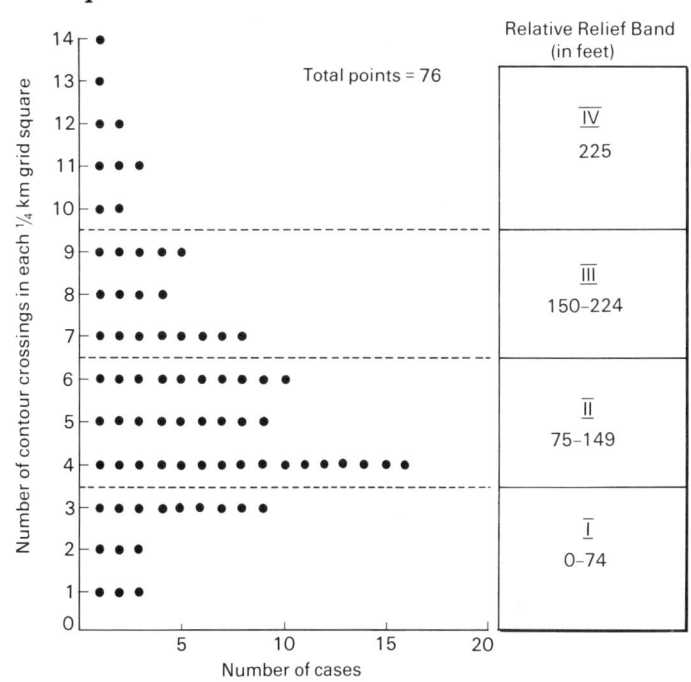

Fig. 4.2 Scattergraph of contour crossings in each ¼ km grid square

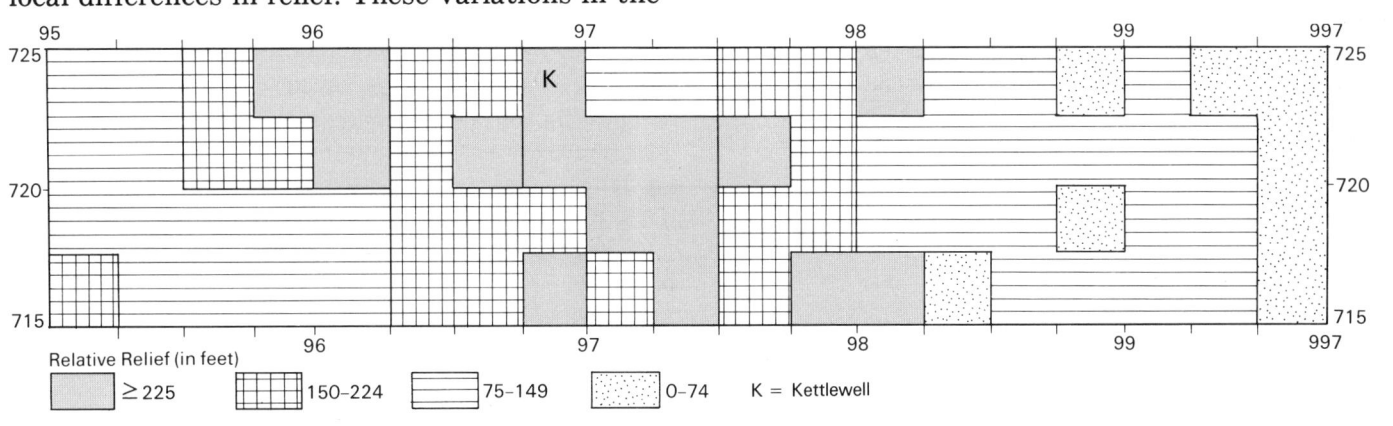

Fig. 4.1 Relative relief diagram (calculation based on ¼ km grid squares)

4. These values are grouped into four bands. The first, 0–74 feet, includes those squares with up to three contour crossings, the second, 75–149 feet, includes squares with between four and six contour crossings, and so on. Examples of these calculations are shown in Fig. 4.3.

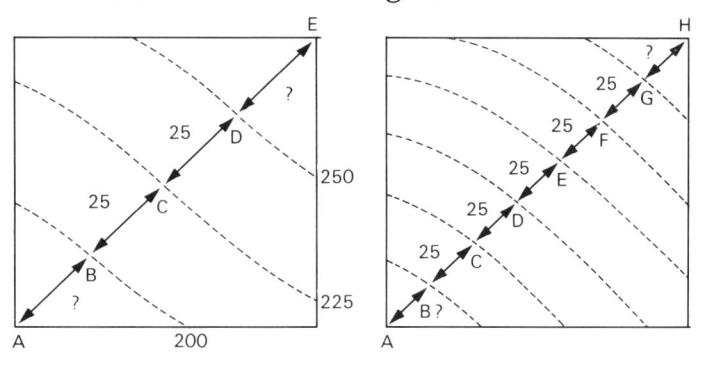

Relative Relief:
AB = ?
DE = ?
BC = 25
CD = 25

Relative Relief: 50–74
Band I

Relative Relief:
AB = ?
BC = 25
DE = 25
FG = 25
GH = ?
CD = 25
EF = 25

Relative Relief = 125–149
Band II

Note, however, that these values calculated for relative relief always underestimate the true values. Actual values cannot be calculated without information for those areas labelled ? on the diagrams.

Fig. 4.3 Calculation of relative relief (all contour values are in feet)

What does Fig. 4.1 tell us about the physical landscape of Wharfedale? Compare the Ordnance Survey map and the relative relief diagram, and then answer these questions.

4.1 Where are the areas with the smallest values for relative relief? What is their geology? (Refer to Fig. 4.4.) Give map references for these areas.

4.2 Describe the relative relief of the valley floor of the River Wharfe, downstream from Kettlewell.

4.3 Which areas have the greatest values for relative relief?

4.4 What landforms are often associated with abrupt changes in the relative relief of the land surface?

4.5 Compare the line section drawn along Northing 72 (Fig. 4.5) with the corresponding part of Fig. 4.1. What conclusions can you draw?

4A.2 · Geology and relief

Information taken from maps of the solid and drift geology of a region is valuable when interpreting the physical landscape. Figure 4.4, a simplified geological map, shows the existence of three main rock types in the Wharfedale area. In order of increasing age, these are:

(i) *Millstone Grit*: a very resistant, coarse, insoluble sandstone rock. It is consolidated in layers which are jointed.

(ii) *The Yoredale Series*: an alternating series of highly permeable limestone beds, and thinner beds of less permeable shales.

(iii) *Carboniferous Limestone*: a very permeable rock underlying the floor of the Wharfe valley. This rock is overlain by deposits from a glacial lake and alluvial fans.

4.6 Compare the rock outcrops on Fig. 4.4 with the contour pattern on the Ordnance Survey map. To what extent can these outcrops be identified in the landscape?

4A.3 · Limestone landscape

The kinds of surface feature which can indicate the existence of limestone strata are described in Chapter 1 (1B.2). With these guidelines in mind, examine the map extract (Map 4), and complete the following exercises.

4.7 Draw a line section along Northing 71, between Eastings 950 and 998 (see Chapter 1, Section 1B.3, for guidance). Use neat, vertical arrows of different length to point to the presence of a scar, limestone pavement, spring, wood, and road on the western side of the valley. Mark these same features on the eastern side, together with the pot hole lying just north of the section line. Indicate the position of the River Wharfe. Add a title to your section, insert both the vertical and horizontal scales, and indicate the direction along which the section is drawn.

On the straight edge which you used to mark the position of contours, insert the boundaries of the main rock outcrops from the geological map. Assume that these strata are bedded horizontally, and mark them on the line section. Add a full key to this diagram.

4.8 Find three places on the map where streams are represented as broken lines. What term is used to describe such a drainage pattern and how does it occur? Illustrate this drainage feature by a sketch map.

4.9 Examine the pattern of springs in grid squares 9674 and 9671. Why do many of these springs tend to occur in lines? What are the altitudes of these lines? What different kinds of rocks are present in these areas? How do these produce the lines of springs?

4.10 Limestone areas are usually characterised by a lack of surface drainage due to the solution and permeability of the rock. Why, therefore, are so many streams shown on the map?

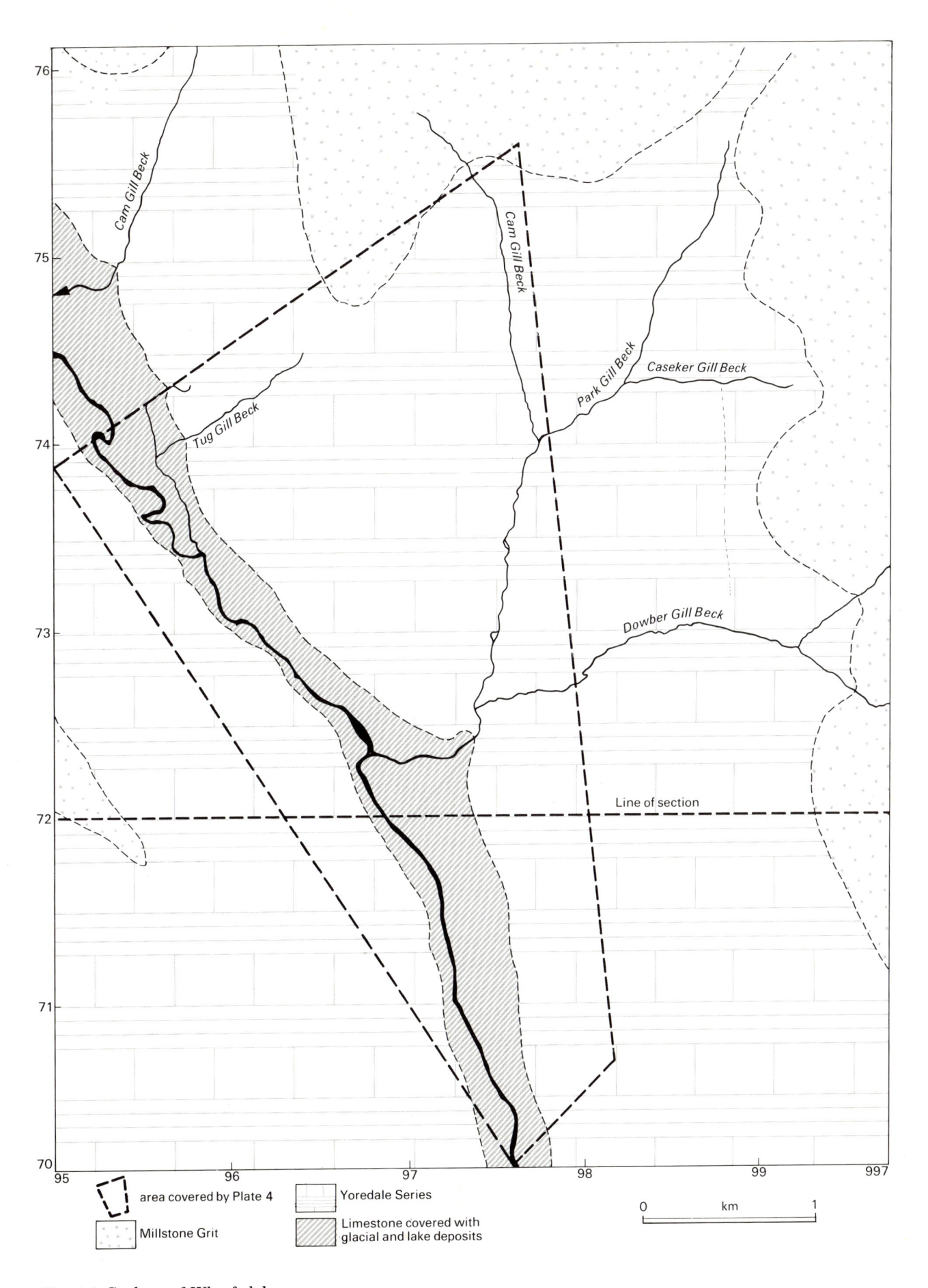

Fig. 4.4 Geology of Wharfedale

4.11 Compare the scenery of the Yoredale Series with that of the Millstone Grit with regard to: (i) surface landforms, (ii) the reaction of rock to running water, (iii) vegetation, (iv) land use.

4.12 Study the valley of the Dowber Gill Beck.
Draw a cross profile from 985726 (the corner of the field) to the spring at 986735. Compare this cross profile with that of the Wharfe Valley prepared for exercise 4.7. If the Dowber Valley is described as V-shaped, how would you describe the Wharfe Valley? Can you explain why these cross profiles are so different?

4.13 What evidence is shown on Plate 4 to confirm the presence of limestone in the landscape? Relate the area covered by the air photograph to the geological detail shown on Fig. 4.4.

4.14 Refer to Plate 4.
(i) Name the valley A.
(ii) Identify and name the scar B.

4A.4 · Transect diagram

A transect diagram can be used to demonstrate the general relationship between features in the physical and human landscapes. Figure 4.5, drawn along Northing 72, covers a representative tract of countryside in Wharfedale. An annotated line section forms the base of the diagram. (Where possible, too, the solid geology can be inserted on this section.) Above this section, at appropriate points, observations are recorded for other features in the physical and human landscapes.

Figure 4.5 was drawn using observations for a band of the countryside half a kilometre to either side of Northing 72. It includes symbols, instead of text, to allow a greater amount of detail to be

Plate 4

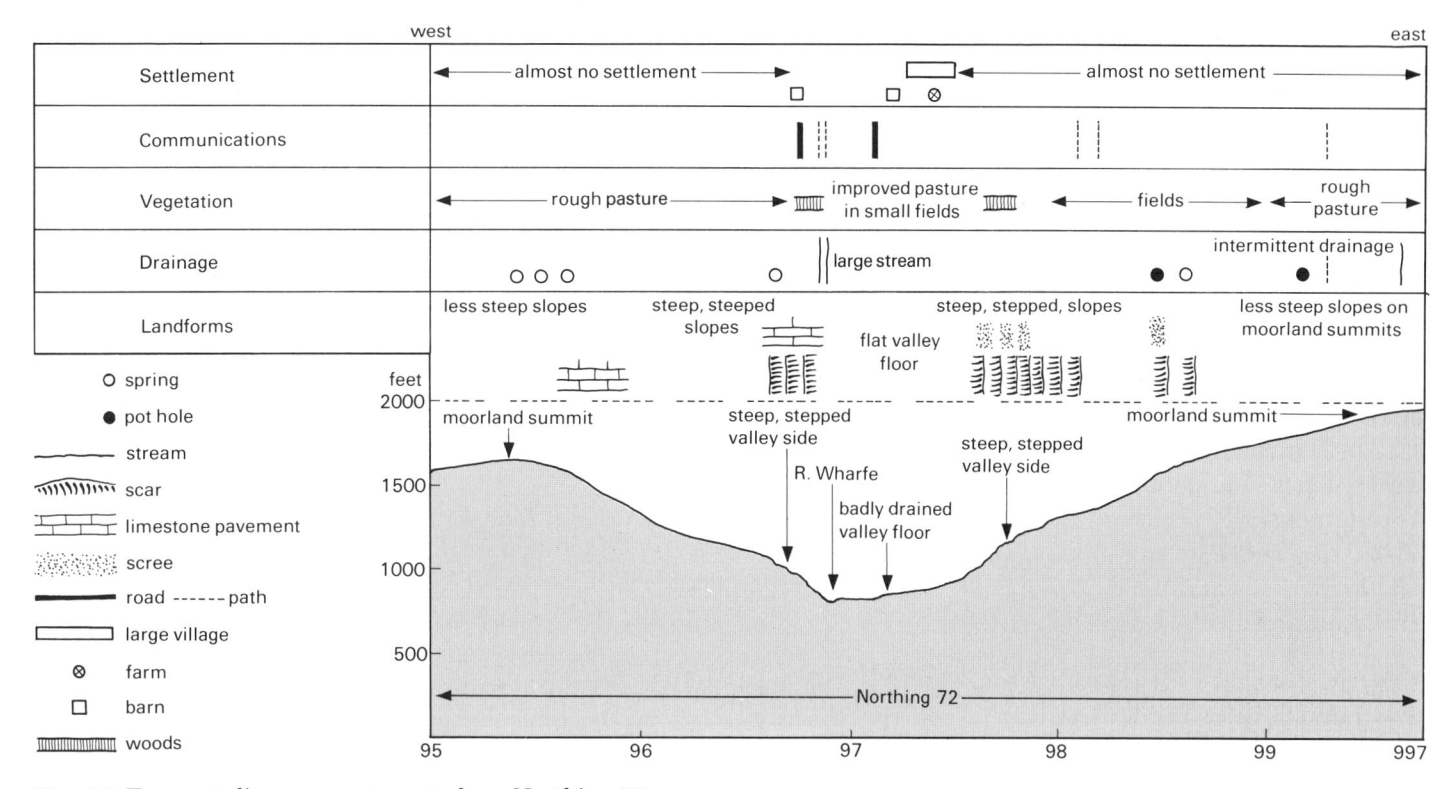

Fig. 4.5 Transect diagram west–east along Northing 72

included. Using the transect we can therefore:

(i) relate individual features in the human landscape to relief and drainage e.g. settlement and relief; communications and relief;

(ii) study the interrelationship between several features at any point on the transect e.g. relief, communications and settlement; landforms, drainage and vegetation.

Study Fig. 4.5, Plate 4 and Map 4, and answer these questions on the distribution of vegetation in Wharfedale.

4.15 Give a map reference for woodland C on Plate 4.

4.16 Prepare a tracing overlay of the woodland distribution from the Ordnance Survey map. Draw selected contours on this overlay. Within what range of altitude does the woodland lie?

4.17 Explain why woodland is not found at higher altitudes.

4.18 What types of trees are most commonly found in this region? Which types are best suited to limestone environments – oak, ash, beech or poplar?

4.19 In the historic past, there was more woodland at lower altitudes in Wharfedale than today. Give three possible reasons why this woodland has been cleared.

4.20 Relate the pattern and direction of field boundaries to the cross profile of the Wharfe Valley (see Plate 4).

4.21 What type of vegetation/land use is represented by the symbol showing faint curved lines of grey dots? (This symbol occurs widely over most of the higher regions of map extract.)

4.22 What is peat? How is it formed?

4.23 Write a short account of the relationship between vegetation and land use, and relief, as shown on Fig. 4.5.

4B · Human activity in the upland environment

4B.1 · Economic activity

Three kinds of economic activity can be identified from the map: agriculture, extractive industries and tourism.

Agriculture. This account of farming in Wharfedale is based on the guidelines proposed in Chapter 1, Section 1D.3. Given the physical characteristics of the region, namely high altitude, steep and often precipitous slopes, thin soils and exposed fields, there are limited opportunities for arable cultivation. Furthermore, there is little flat land on the floors of the V-shaped tributary valleys while the wider floodplain of the Wharfe may be prone to seasonal flooding as indicated by the existence of artificial embankments along sections of the river bank (9574; 9573).

The higher regions of the valley sides, mainly on Millstone Grit, are covered with rough pasture, heather and peat. These areas probably support sheep farming. In winter the flocks will be driven down the valley side to the sheltered situations on the lower slopes or valley floor. Some flocks may be transported by road from Wharfedale to a lower-lying region with a less demanding climate. The several barns recorded at lower altitudes provide shelter for livestock and storage for winter fodder.

It is possible, too, that cattle are reared on the lower and more gentle slopes. In such situations, the farmers would need to improve the quality of the land and drain lower-lying fields.

4.24 Compare the farming activities in Wharfedale with those described for the Nant Ffrancon region in Chapter 3 (3C.1). How similar are these physical environments for agriculture?

Extractive industries. There is historical evidence in the landscape of a mining industry: for example, old shafts (982759; 988706; 967736 – 978757) and a disused mine (993728). The kinds of material mined, however, cannot be determined, although it is known that in other parts of Britain lead and iron ore deposits are sometimes associated with limestone rocks. Quarrying for road metal and building stone (for housing and dry-stone walls) was also of local significance (969745).

4.25 Compare and contrast the distribution and scale of mining activity in Wharfedale with that for quarrying in Snowdonia (Map 3).

Tourism. Examine the map and read the comments on tourism and the physical environment in Chapter 3, Section 3C.2. Now answer the following questions.

4.26 What evidence is there on the map to suggest the presence of a tourist industry?

4.27 As the publicity officer for the local authority, you have to prepare a tourist information leaflet on Wharfedale. Refer to the O.S. map and Plate 4 and write a short statement on the attractions of this area for those interested in spending a walking holiday based at Kettlewell.

4.28 A friend with a strong interest in outdoor pursuits has written to you seeking your advice on a holiday location. He wishes to include rock climbing, pony trekking and underground exploration. Prepare a short reply, comparing the holiday opportunities afforded by the physical environments in Wharfedale and Snowdonia (Map 3).

4.29 You are leading a field party to examine the local landscape. The proposed route is as follows: leave your car parked at 967723; cross the river, turn right past the post office to the chapel (974724); turn right and follow the pathway marked by a double broken line on the map to 993727; again turn right, leaving the path, and walk south along the top of the fields to 997709; join the path here and return to the valley, passing the old shafts at 988706 and rejoining the minor road at 976708; return along minor road to Kettlewell.
 (i) How long is this walk in kilometres? Estimate the time needed for this excursion, including two hours to view points of interest and for drawing field sketches and diagrams, and half an hour for lunch.
 (ii) Identify, with map references, six features of interest to geographers which you would encounter on this walk.
 (iii) What clothing would you prescribe for such a walk in July? What kinds and scales of maps would you advise to assist with the interpretation of the landscape on this route?

4B.2 · Settlement patterns and society

Throughout history the physical environment has exercised a control on the distribution and kind of settlement in Wharfedale. To investigate this relationship in more depth, prepare answers to the following questions, using the map, related diagrams and Plate 4.

Historical settlement

4.30 Which of the following environments were best suited to the economies of primitive peoples?
 (i) The poorly drained and wooded, but warmer, valley floor.
 (ii) The relatively drier, more exposed, cooler grassy slopes of Yoredale limestone at about 1000 feet.
 (iii) At a higher altitude, the peat and heather covered Millstone Grit moorlands.
Justify your answer, with reference to the distribution of historical settlement, the availability of water supply and the geology.

4.31 What are the highest and lowest altitudes at which there is surviving evidence of the historical occupation of this region? Explain why the floor of the Wharfe valley shows little evidence of historical settlement.

4.32 What do the symbols in the south-east corner of grid square 9571 represent?

Contemporary settlement. Chapter 9 (9C.1), provides guidelines for the study of settlement patterns. Refer to these, and then answer the questions below.

4.33 Name the three largest nucleated settlements. Which one has the most churches? What public services does this settlement possess that the others do not?

4.34 Why are these larger settlements situated in the main valleys?

4.35 What is the highest altitude at which present settlement is found? Why do you think that West Scale Park (977746) and East Scale Park (987747) are sited on the same sides of their respective valleys?

4.36 What is meant by the name 'barn' e.g. Low Holme Barn (973707), Rose Well Barn (954704)?

4.37 Study Plate 4 and comment on the changes in the settlement pattern at Scargill House (976712).

4.38 Name, with map references, the settlement units D, E, F and G on Plate 4.

Think about the characteristics of the population living in the mountainous regions of Britain and answer these questions:

4.39 Do you think that the number of people living in the valleys shown on the map is increasing or falling?

4.40 Do you think that the average age of the resident population is greater now than it was fifty years ago? If so, what has happened to the younger age groups in society?

4.41 What kinds of people might come to live in the valleys and thereby reduce the decline in population numbers?

4.42 Compare your answers to questions 4.39 – 4.41 with those you might give for a similar study of Nant Ffrancon (Map 3).

A geographical study of Starbotton. Describe the site, situation, form and functions of Starbotton. Base a short geographical account on your answers to questions 4.43 and 4.44. First of all, however, read carefully Section 9C.1.

4.43 Draw a sketch map, at twice the map scale, of the area enclosed by 950753, 960753, 950743 and 960743. Mark and name the following features:
   (i) contours for 700, 725, 750, 800 and 900 feet; use symbols to represent other important physical features such as scars and screes; add a key;
   (ii) the River Wharfe, Cam Gill Beck and other features of the local drainage system;
   (iii) roads and building outlines.

4.44 Draw a line section between 960750 and 950747. Use a vertical scale of 1 inch to 1000 feet. Print clearly the title and scales. Mark the position of the road and the limits of the built-up area. Now answer these questions:
   (i) Between what contours does the settlement lie? What kind of landform does it cover? What processes produced this landform and how is its origin connected to the processes of glaciation and river erosion?
   (ii) Why was Starbotton built on this landform?
   (iii) Why do you think that the village does not extend along the valley of the Cam Gill Beck?
   (iv) What natural drainage flows through the village? Has this anything to do with the original site of the settlement? Why was the village built above the floor of the main valley?
   (v) Is there evidence to show the recent growth of Starbotton? If so, where have extensions taken place?
   (vi) Above, the word 'village' has been used to describe Starbotton. Do you agree with this description?
   (vii) Is Starbotton a parish in its own right or is it linked with another?
   (viii) What public services are available in Starbotton? Where is the nearest school?
   (ix) Would you describe Starbotton as a route centre? For comparison, refer to Kettlewell.
   (x) Is there evidence of mining and quarrying in the immediate vicinity of Starbotton?
   (xi) What employment is available to those of working age who live in Starbotton?

CHAPTER 5
South Devon

Chapter Plan

5A · **Coastal landforms**
    5A.1 Rock type and structure
    5A.2 Physical regions
    5A.3 Structural influences and river capture
    5A.4 Changes in sea level and landform development
    5A.5 Coastal landforms

5B · **Settlement and the physical environment**
    5B.1 The Kingsbridge Estuary: historical setting
    5B.2 Urban settlement: changes in function
    5B.3 Rural settlement: setting, pattern and economy

5C · **Communications**
    5C.1 The transport network
    5C.2 Categories of road in South Devon
    5C.3 Route description: general guidelines
    5C.4 Topology and the analysis of route networks

5A · Coastal landforms

Three main themes are examined in this study of South Devon (Map 5). Section 5A, which analyses the physical geography of the area, defines the main relief regions and examines the processes involved in shaping the coastline. Section 5B relates to this physical background and considers changes in the human geography of the area, with particular reference to the settlement pattern. Finally, Section 5C introduces simple techniques for the measurement and description of communication networks and applies these to South Devon.

5A.1 · Rock type and structure

Several factors control the shape of a coastline. These include:

(i) the relief of the land surface, relative resistance of the underlying rocks, and lines of structural weakness;

(ii) in the context of geological time, the emergence or submergence of the land surface relative to the level of the sea;

(iii) the gradient of the sea floor immediately offshore;

(iv) the trend of the coastline in relation to the dominant winds and tidal movements;

(v) the processes of erosion and deposition in creating or modifying coastal landforms.

There are three main structural trends in the Kingsbridge region. These are shown by the orientation of major sections of the coastline, and the alignment of the main river valleys. The trend lines, shown diagrammatically on Fig. 5.1, are:

North–South (......): parallel to the main axis of the Kingsbridge Estuary and including major sections of both the east-facing and west-facing coastlines.

North East–South West (———): parallel to the coastline between Start Point (8337) and Prawle Point (7735).

North West–South East (—·—·—): parallel to the coast between Bolt Tail (6639) and Bolt Head (7235).

Rivers and marine erosion have exploited these lines of structural weakness, giving rise to the distinctive trend in the physical landscape.

5A.2 · Physical regions

Three distinctive regions in the physical landscape can be delimited using the guidelines in Chapter 1, Section 1D.1. The general limits of each region are indicated on Fig. 5.1.

Southern dissected plateau. This region lies south of Northing 39 and extends to either side of the Kingsbridge Estuary. Its surface characteristics can be summarised by preparing a *contour trace, representative line section* and *relative relief diagram*. (Refer to Chapter 1, Section 1B.3.)

The contour trace, superimposed on Fig. 5.1, shows that the plateau surface west of the estuary averages 131–133 m in altitude; to the east it is a little higher, with altitudes up to 138 m. Along the coast there is an alternating sequence of headlands and bays or coves. The four dominant headlands, Bolt Tail (6639), Bolt Head (7235), Prawle Point (7735) and Start Point (8337), indicate the presence of more resistant strata. The plateau surface above 120 m is

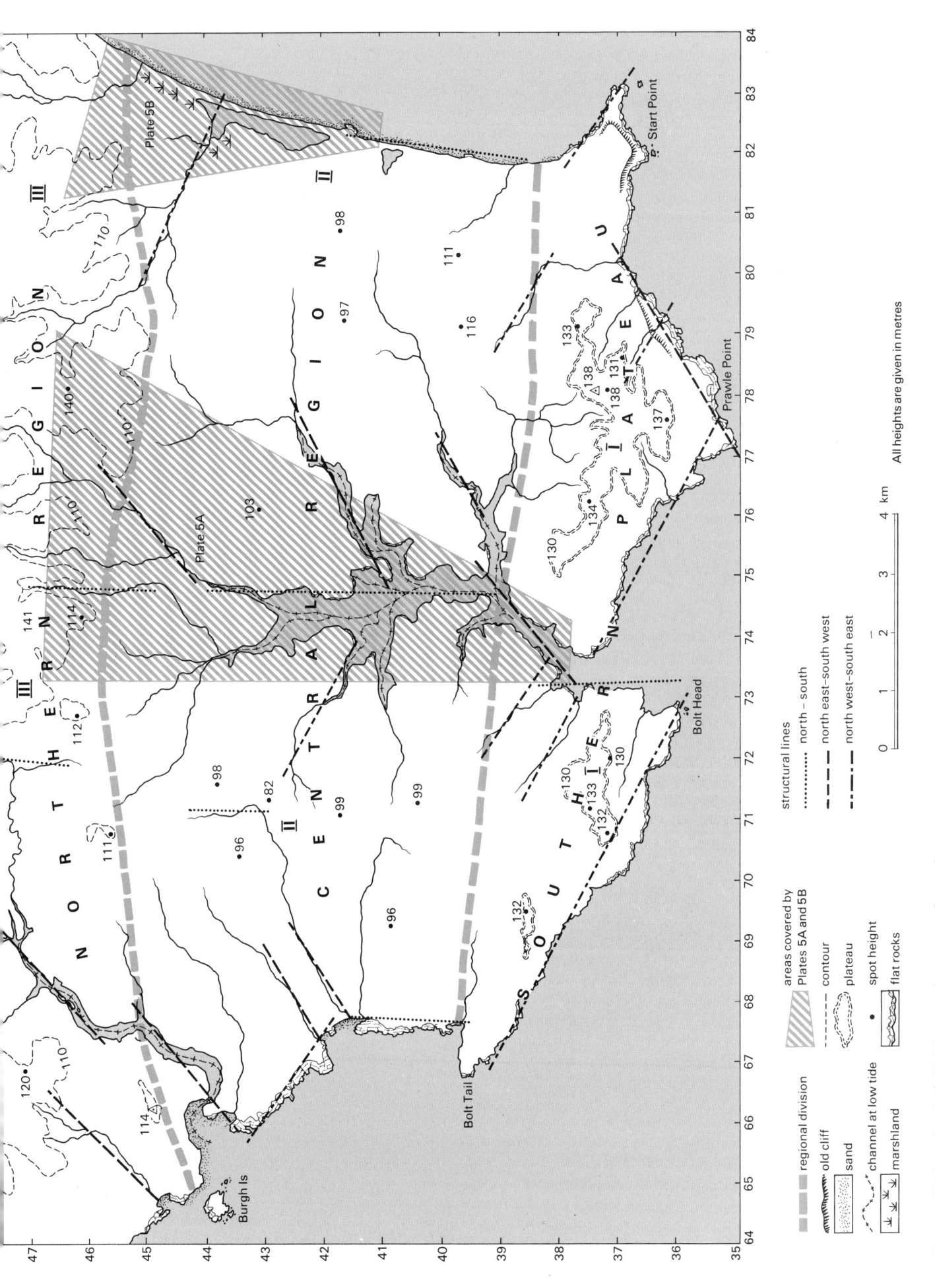

Fig. 5.1 Major physical divisions and structural trends

exposed to south-westerly winds and is almost devoid of woodland and settlement. The edges of the plateau surface are deeply dissected by small winding streams. In direction, these normally conform to one of the major trend lines. Their valleys are typically narrow, lack a flat floor, and have steep sides with interlocking spurs. Such features are typical of the upper course of a stream (see page 21).

5.1 To complement this general description, prepare the following diagrams.
 (i) An *annotated line section* from Shoelodge Reef (823376) to the cove at Soar Mill (697376).
 (ii) A *relative relief diagram* between Northings 37 and 38, and Eastings 69 and 83. (Refer to Chapter 4, 4A.1, for guidance in constructing this diagram on a ¼ km grid.)
 Explain how these two diagrams support the contour trace and help with the interpretation of the physical landscape.

Central region. To the north of the plateau, between Northings 39 and 44, the land surface undulates and rarely exceeds 100 m in altitude. In this region the valleys are more open, and have less steep sides. This may indicate that they have developed in less resistant rocks. Marshland covers the flat floors of two open valleys to the west of Easting 70 (6742; 6842).

North region. To the north of Northing 44, the altitude of the land surface increases gradually to 150 m. The general southern limit of this region is shown on Fig. 5.1 by the 100 m contour. Several steep-sided valleys dissect the landscape. The altitude of the land, and typical cross profiles of the numerous valleys, would suggest that the underlying strata are more resistant than those in the Central Region.

5.2 Draw annotated line sections between (i) 685390 and 685460 (ii) 700465 and 835465.
 Compare the cross profiles of the valleys. Do these line sections illustrate the contrasts in land surface between Region II and Region III? Give reasons to support your decision.

5A.3 · Structural influences and river capture

The relationship between lines of structural weakness and the direction of flow of the main rivers in the drainage system has been discussed above. As Fig. 5.2 shows, the headwaters of the Kingsbridge Estuary were captured by the River Avon which has cut back along a line of

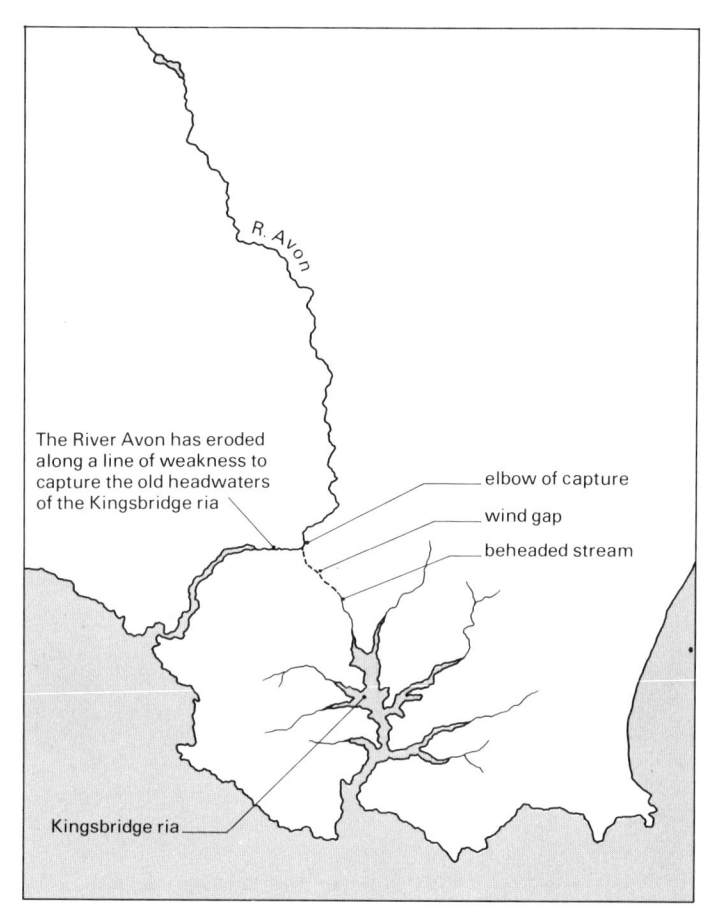

Fig. 5.2 River capture on the River Avon (diagram extracted from O.S. 1:50 000, Sheet 202)

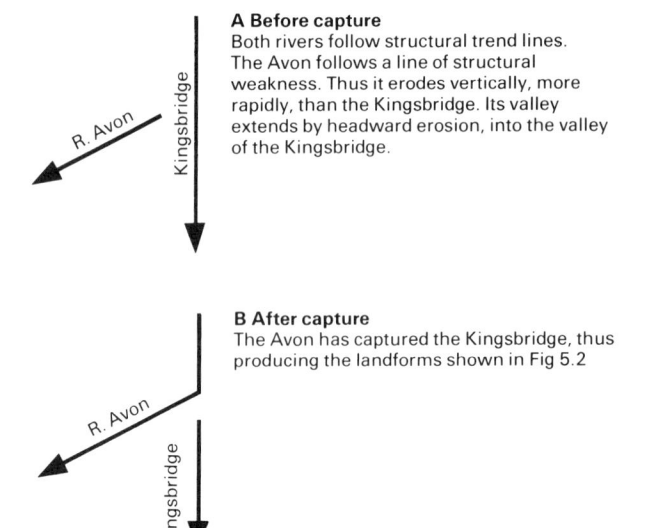

Fig. 5.3 Explanation of river capture

A Before capture
Both rivers follow structural trend lines. The Avon follows a line of structural weakness. Thus it erodes vertically, more rapidly, than the Kingsbridge. Its valley extends by headward erosion, into the valley of the Kingsbridge.

B After capture
The Avon has captured the Kingsbridge, thus producing the landforms shown in Fig 5.2

structural weakness. This process of river capture has left a wind gap in grid square 7245.

5.3 Examine the drainage system and contour pattern in the area 700440/700475 – 740440/740475. Refer to Fig. 5.3 and reconstruct the process of river capture and its related landforms.

Map 4 Wharfedale 1:25 000

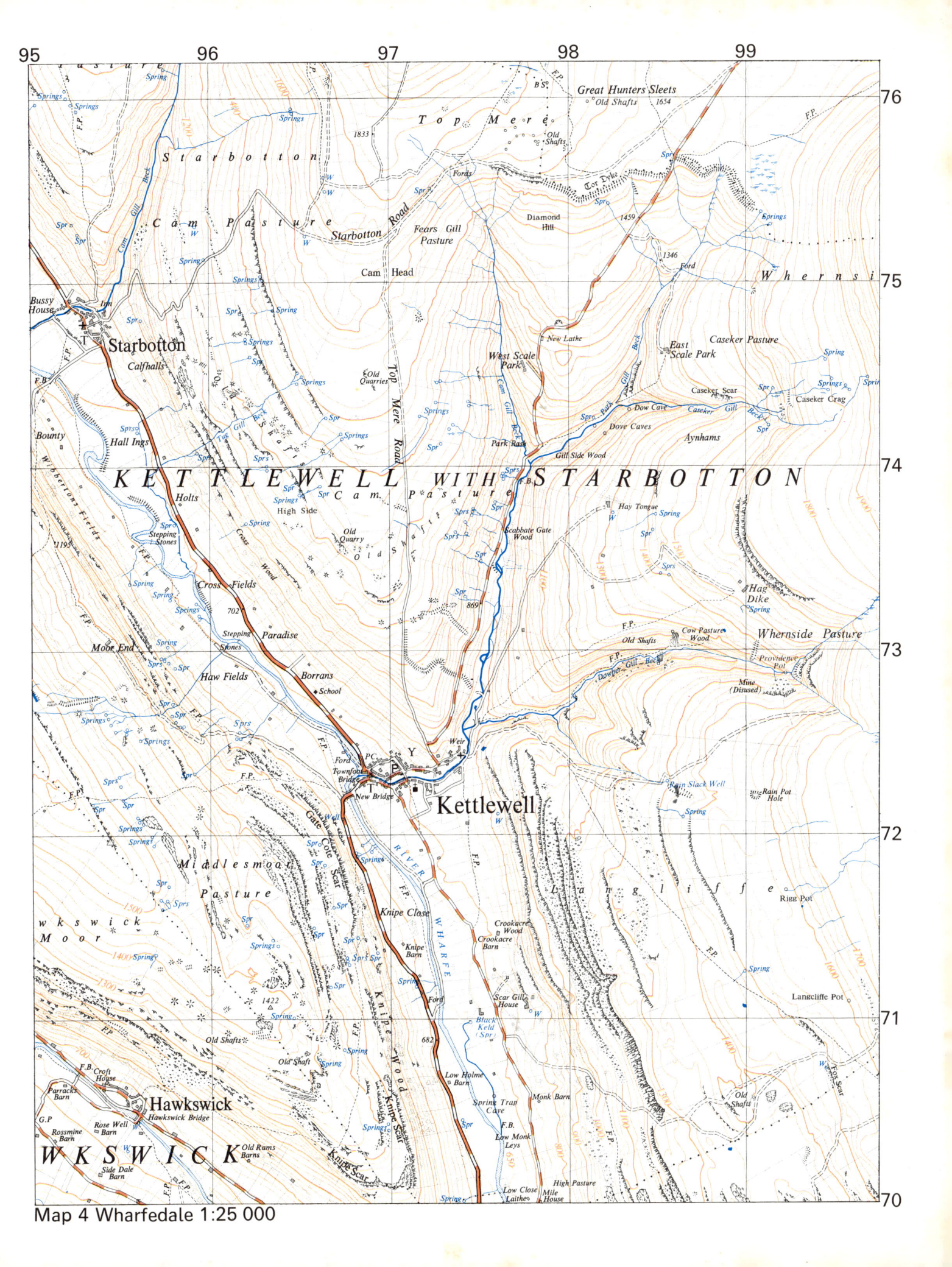

Map 5 South Devon   1:50 000

Knighton
Cross
Norton
Buckland
Merrifield
Strete
Ledstone
Goveton
Flear Fm
Cole's Cross
Dittiscombe
Colehanger
Dearswell
Whitestone Cross
Loworthy
Rimpston
Kingsbridgefork Cross
Alston
Pittaford
Netherton
Pinhey's Cross
Blackland
Homelands
Buckland-tout-Saints (Hotel)
Centry
Resr
Valley Springs
Higher Coltscombe
Lower Green Cross
Strete Gate
Fursdon
Battle Ford
Poole
Malston Barton
Stancombe Cross
Harleston
Start
Higher Ley
College
Field Study Centre
Ranscombe
Marepark Cross
Slapton
Bearscombe
Darnacombe
Mon
Slapton Sands
Nature Trail
Bowcombe
Sherford Down
Sherford
Homefield
The Castle
Fort
Bowden
Slapton Ley
SBRIDGE
Duncombe Cross
Coleridge Ho
Frittiscombe
France Wood
Nature Reserve
Lower Ley
The Grange
Keynedon Barton
Coombe Park
Chillington
Stokenham
Cemy
East Charleton
Frogmore
Hotel
Inn
Sch
Carehouse Cross
A 379
West Charleton
Mill Fm
Torcross
Widewell
Loo Cross
Winslade
Molescombe
Mattiscombe
Cross
Hotel
Limpet Rocks
Frogmore Creek
Ham Point
Widdicombe Ho
Cotmore
North Pool
Kernborough
Sunnydate
BRIDGE
UARY
Wareham Point
Halwell Fm
Beeson
Moyson
Beesands
Lower Barn
South Pool
Dunstone
Ford
Huckham
Combe
Tosnos Point
Middlecombe Fm
Scoble
Cousin's Cross
Tinsey Head
Westerncombe
Gullet Fm
Batton
Southpool Creek
Devon South Coast Path
Wilton
Kellaton
LB Sta
Muckwell
Greenstraight
Ford
Goodshelter
Hallsands
East Portlemouth
Waterhead
Chivelstone
South Allington
Bickerton
Hotel
Mill Bay
Hollowcombe Head
Holset
Hare Stone
Down Fm
West Prawle
Rickham
Borough
Shoelodge Reef
Nestley Point
Start Fm
Portlemouth Down
Moor Fm
Wr Twr
Woodcombe
Lannacombe Beach
Gara Rock
Hotel
START P
The Narrows
Deckler's Cliff
East Prawle
Two Stones
Black Stone
Ravens Cove
Shag Rock
Sleaden Rocks
Peartree Point
Ballsaddle Rock
Lannacombe Bay
Pig's Nose
Gorah Run
Gorah Rocks
Ham Stone
Sharpers Head
Ball Rock
Devon South Coast Path
Gammon Head
Brimpool Rocks
CG Sta
Langerstone Point

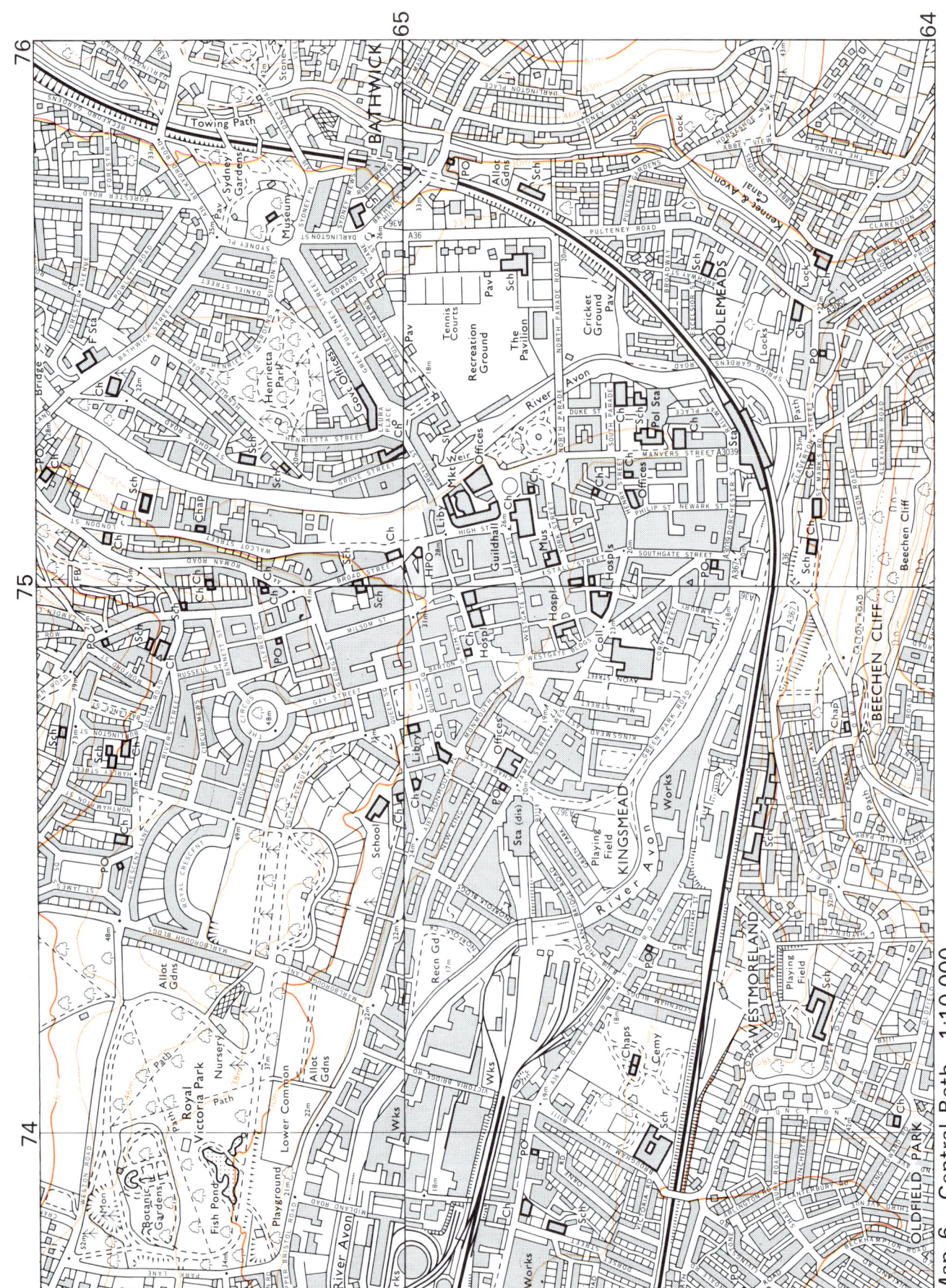

Map 6 Central Bath 1:10 000

5A.4 · Changes in sea level and landform development

In recent geological times the relative levels of the land and sea have changed. This process has produced two contrasting types of coastline in Britain: *emergent* coastlines and *submergent* coastlines. Features of both kinds of coastline are evident in the landscape of South Devon.

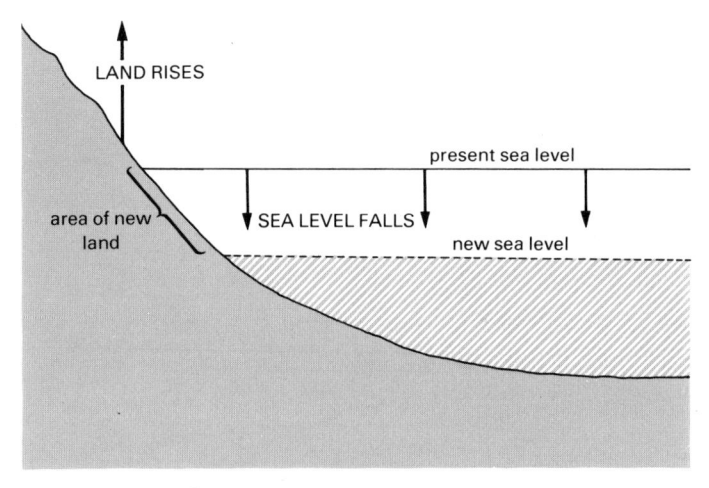

Fig. 5.4 Coastal emergence

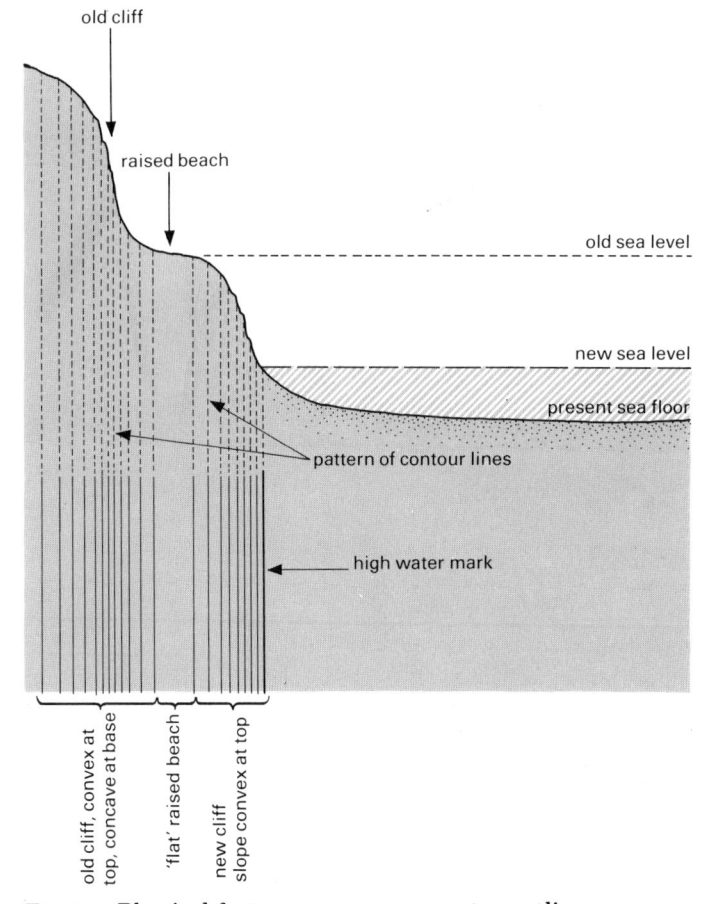

Fig. 5.5 Physical features on an emergent coastline

Emergent coastlines. Figure 5.4 shows that as the sea retreats, areas of new land are created. Raised beaches are a typical result of this process. Examine the pattern and spacing of contours up to one kilometre inland from Prawle Point (7735) and Start Point (8237/8236). Relate your observations to Fig. 5.5 which shows some of the features normally associated with emergent coastlines. In some circumstances more extensive marine platforms are formed, like that to the south of Northing 39.

5.4 Why is the surface of the plateau (Region I) so flat? Do you think that river erosion would have produced such an extensive and flat surface? If not, examine the likelihood of other possible causes.

5.5 Examine the river valleys around the edges of this plateau. Can you explain their steep sides and narrow width? (Consider, for example, the importance of changes in the relative levels of land and sea and the nature of the underlying rocks.)

Submergent coastlines. When sea level rises relative to the land, the lower courses of existing rivers become flooded (Fig. 5.6). This process is clearly illustrated by the large estuary between Kingsbridge and Salcombe, and the lower reaches of the River Avon, downriver from Bridge End (6946). These drowned estuaries or rias are characteristic of submerged coastlines. Notice how the original drainage network of the main river and its tributaries is preserved on the floor of the ria (Fig. 5.1).

5.6 Why are there extensive areas of tidal flats along the main channel and branches of the ria between Kingsbridge and Salcombe?

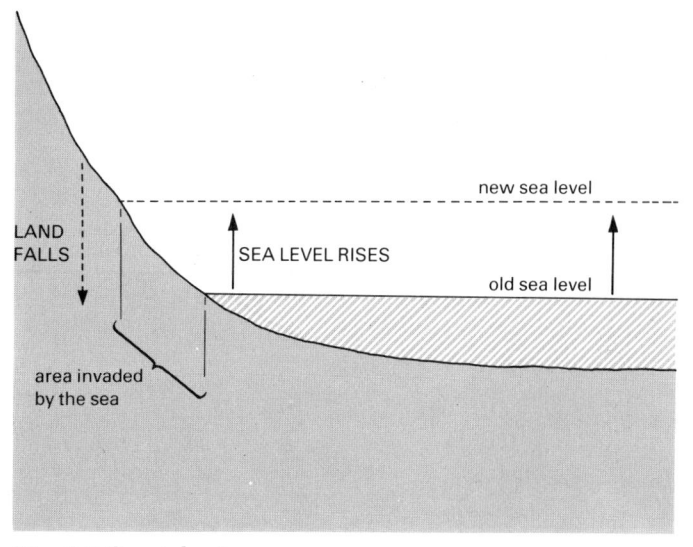

Fig. 5.6 Coastal submergence

Plate 5A

Examine the map and Plate 5A, and answer these questions on the Kingsbridge Estuary.

5.7 What proportion of the surface of the estuary is uncovered by water at low tide?
 (i) less than 25%
 (ii) 25%–50%
 (iii) 50%–75%
 (iv) more than 75%

5.8 Examine Plate 5A. Was the tide in or out when this photograph was taken? Give reasons to support your answer.

5.9 Compare and contrast the sections of coastline between the following points: 745423–760417; 739423–746394; 669422–666397.
 What are the major differences between these sections? What marine processes are active in shaping these shorelines?

5.10 Give map references for the areas of woodland A, B and C on Plate 5A. What is common to these situations?

5.11 Explain the difficulties an inexperienced yachtsman might face when sailing a small boat from South Sands (7237) to Kingsbridge in the following conditions:
(i) with an ebbing tide (ii) at high tide.

5A.5 · Coastal landforms

To interpret the local variety in coastal landforms it is important to understand how the processes of marine erosion and deposition relate to the land surface and geological structures above the high water mark.

Headlands and bays. Both structural weaknesses and less resistant types of rock are exploited by the processes of marine erosion. Areas of more resistant rock and stronger geological formations, therefore, remain as peninsulas, promontories or headlands on the coastline, while the weaker areas are eroded to form coves and bays.

Examine the coastline between Butter Cove (6643) and Bolt Head (7235) and note the alternation of headlands and coves. This stretch of the coast contains a large headland, Bolt Tail (6639), together with smaller ones like Warren Point (6642) and Thurleston Rock (6741). In such situations it is common for beaches to develop at the heads of small bays (6741; 6739). These beaches comprise the fine material which is eroded from the headlands and transported by wave action and tidal movements. This sand is deposited in the shallower, slacker water at the head of the bay.

Islands, stacks and tombolos. From map evidence alone it is impossible to determine whether Burgh Island (648438) was severed from the mainland by erosion or whether it became an island as a result of the general processes of submergence. There is a considerable deposit of sand and mud on its landward side. It therefore seems possible that in the future a tombolo, a type of sandspit, will join this island to the mainland.

Stacks are much smaller features and there are several examples along the coastline, for example Ballsaddle Rock (797365) and Hare Stone (820380).

Bay bars, barrier lakes, truncated spurs and dead cliffs. Longshore drift was responsible for the movement of sand which forms a bar across the mouth of the minor stream immediately north of Beesands (8240). The completion of this constriction led ultimately to the development of a lake (Fig. 5.7). Farther north, between grid squares 8242 and 8345, there is further evidence of the deposition of beach materials by longshore

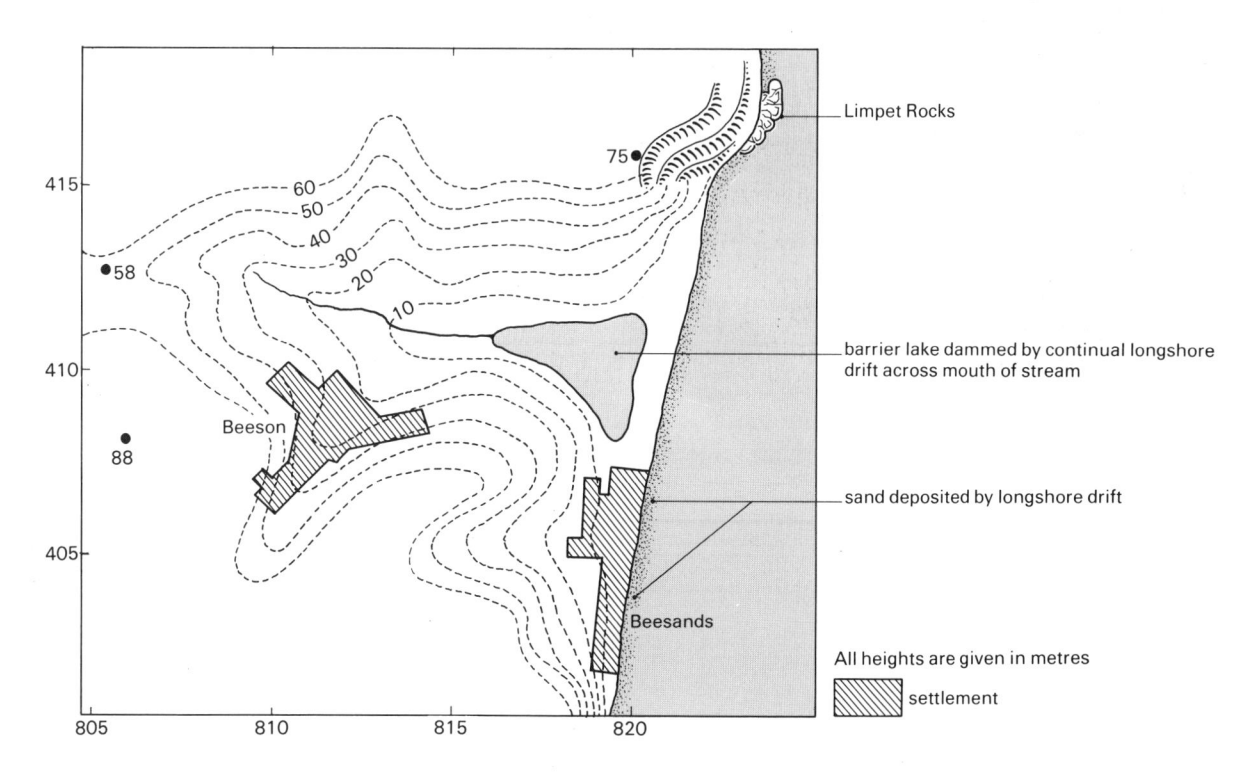

Fig. 5.7 Sketch map of Beesands

Plate 5B

drift. Part of this feature is shown on Plate 5B (A–B). Unfortunately, the direction of the drift cannot be determined from the map. But since it is usually governed by the prevailing winds coming from the most open stretch of water, you may be able to deduce the direction.

Bay bars have thus protected the earlier line of sea cliffs from recent marine erosion (Plate 5B). As a result, the long spurs of higher land, trending east–west, terminate in very steep slopes to form dead cliffs, as shown in 8243 (see Plate 5B, X–Y).

There are no spits or bars along the west coast, which is more irregular in shape. Here, the headlands are exposed to the dominant westerly winds and, apart from bay-head beaches (6741), there is little evidence of recent deposition, the waves removing eroded material.

5.12 Describe the physical processes which are shaping the coastline in the following locations on Plate 5B: A–B, C, D, E. Which of these locations show evidence of (i) erosion (ii) deposition?

5.13 Give, with grid references, three examples of bay-head beaches between Bolt Tail (6639) and Start Point (8337).

5.14 Name, with map references, three examples of

58

headlands with detached stacks between Bolt Tail and Start Point.

5.15 Examine the coastline at Hove on Plate 1B. What process of beach development can be seen? From which direction are the waves transporting beach materials?

5.16 Why is the coastline between Beesands (819405) and Slapton Sands (830445) attractive to tourists? List evidence from the map and Plate 5B to support your answer.

5.17 Assume that the South Devon Coastal Path is continuous between 6642 and 8242. Prepare a plan for a walking holiday which uses this route between Thurleston (6742) and Torcross (8242). It is understood that you will carry camping gear and maintain a walking speed of 4 km per hour.
 (i) How long would you take?
 (ii) Where will you spend each night? Will you divert to a regular camping site or ask permission of a farmer? Remember, you are not permitted to camp anywhere you may wish. At which farm(s) would you seek permission to camp overnight?
 (iii) Where would you cross the Kingsbridge Estuary?

5B · Settlement and the physical environment

5B.1 · The Kingsbridge Estuary: historical setting

In South West England, where the lower courses of southward flowing rivers have become drowned to form rias, major settlements have developed at either the ria head or near its mouth. As the map shows, the Kingsbridge Estuary has urban settlements at both locations, namely Kingsbridge and Salcombe.

It seems clear that the Kingsbridge Estuary has had an effect on the development and distribution of settlement, and also the routes of communications in South Devon. It is likely, too, that economic activities connected with the sea, namely fishing, trade and latterly tourism, have broadened the base of the local farming economy and provided a wider choice of employment for the local population.

In the days before there were good quality metalled roads, small vessels plied the coastline of Britain. Some sailed far inland along the estuaries and navigable rivers. Having shallow draughts, these boats provided cheap transport for bulky commodities like coal, building materials, grain and other merchandise needed by local communities. They also carried agricultural produce from rural areas to major urban markets elsewhere in Britain and Western Europe.

5B.2 · Urban settlement: changes in function

In the context of these historical comments and your knowledge of the physical geography of the area, refer to Chapter 9, 9C.1, and prepare answers to the questions on Kingsbridge and Salcombe.

Kingsbridge

5.18 Do you think that Kingsbridge was ever a river port? Prepare a short written statement, based on your answers to the following questions.
 (i) What difficulties might have restricted the growth of Kingsbridge as a port?
 (ii) What advantages of site and situation does Kingsbridge have as a small river port?
 (iii) What locational advantages as a river port does Kingsbridge have in contrast to Frogmore (7742), South Pool (7740), and Salcombe (7439)?

5.19 Which settlement, in your view, had the greater potential for development as a small port: Kingsbridge (7343) or Outer Hope (6740)? Justify your answer with regard to (i) physical site and situation; (ii) regional communication links.

5.20 Use a graticule or grid like that shown in Fig. 3.8 and draw a sketch map showing the site and situation of Kingsbridge. Mark the following features: (i) head of the ria, (ii) 50 m contour, (iii) A381 and A379 main roads, (iv) outline of the built-up area.

5.21 Relate the routes of the main roads serving Kingsbridge to the local topography. Why do you think these routes were chosen?

5.22 To what extent is Kingsbridge a route centre?

5.23 Find the course of the old railway leading north from Kingsbridge. Why was it built? Why was it not continued to Salcombe? Why do you think it was abandoned? What does the existence of this railway tell us about the importance of Kingsbridge?

5.24 What employment is available to the residents of Kingsbridge? Justify your answer with reference to specific land uses shown on the map, and your knowledge of activities which are normally carried on in small towns serving agricultural regions.

5.25 Is there any evidence of manufacturing industries in Kingsbridge?

5.26 Refer to Chapter 7, 7C.1. Identify the historical core of Kingsbridge. Note the form and location of the more recent residential developments. Draw a sketch map to illustrate your observations. How does Plate 5A help with this exercise?

Salcombe

5.27 What physical and economic factors have restricted the development of Salcombe as a major port?

5.28 Study the map and Plate 5A. Draw a simplified sketch to indicate the main growth regions of Salcombe. Relate these to the physical site of the town.

5.29 Read Chapter 6, 6C.2. Contrast the residential environments of regions D and E on Plate 5A with respect to: (i) kinds of houses, (ii) density of development, (iii) aspect and exposure.

5.30 What present-day activities and environmental features attract visitors to Salcombe in (i) winter (ii) summer?

5.31 Imagine that you have to prepare a tourist brochure designed to attract elderly visitors from London. What environmental features of the Salcombe region would you highlight in the text?

5B.3 · Rural settlement: setting, pattern and economy

5.32 Read carefully the following statement on the pattern of rural settlement in South Devon:

> The village economies in South Devon are related more closely to agriculture than to sea fishing. In the south, the plateau surface is almost devoid of nucleated settlement. Elsewhere, the villages and hamlets are found mainly at heads or on the sides of valleys, but always near a supply of fresh water. The network of minor roads servicing these settlements follow routes which minimise the changes in gradient resulting from the widespread dissection of the landscape.

Do you agree with this description? Give reasons to support your answer, quoting appropriate evidence on settlement types and the arrangement of minor roads.

5.33 Name, with map references:
(i) three villages sited at the head of a valley, just below the plateau surface (Region I on Fig. 5.1);
(ii) two villages in the bottom of a valley at the crossing point of a stream;
(iii) two villages on the side of a valley, facing south;
(iv) two coastal villages.
From your answers, what can you say about the relationship between village sites and the physical geography of South Devon?

5.34 How many buildings are located alongside the following routes:
(i) the A381 between West Alvington (7243) and Malborough (7039);
(ii) the most direct minor road linking Chillington (793428) and East Prawle (778367)?
What factors explain these distributions of settlement?

5.35 Find Buckland (6743), South Milton (6942), Osborne Newton (6945) and Torcross (8242). To what extent are these settlements sheltered from the full force of the westerly and south-westerly winds?

5.36 Examine the sites of the following villages and hamlets: South Allington (7938), West Prawle (7637), Rickham (7537), Frittiscombe (8043), Kernborough (7941) and Frogmore (7742). Describe each site with respect to: aspect, shelter, position in the valley or on the plateau surface, proximity to a supply of fresh water.

5.37 What employment do you think is available to the residents of (i) Thurleston (6743), (ii) Slapton (8245), (iii) Outer Hope (6740)? Are these villages conveniently situated for those wishing to travel daily to employment in Kingsbridge?

5.38 Imagine that you have been appointed as a geography teacher in Kingsbridge School (732438). You are prepared to drive not more than 16 km each day, and prefer to live in a village. In which villages would you seek a house? Justify your choice of settlements.

5.39 Assemble evidence from the map and Plates 5A and 5B to demonstrate the importance of the tourist industry to the economy of the Kingsbridge region.

5.40 Is quarrying an important feature in the economy of any village on the map? Give examples to support your answer.

5C · Communications

5C.1 · The transport network

In Britain the network of land communications – roads, railways and canals – has been developed over a long period of time. Factors such as changes in the distribution of population and the development of new industries have been important influences on the spatial arrangement of these networks. Adverse features in the physical environment can readily increase both the costs and difficulties of route construction. Roads and railways, for example, may need tunnels to negotiate a mountainous area or causeways to cross tracts of marshland, while canals may depend on locks to maintain an acceptable gradient.

Some parts of a network which carry large volumes of traffic may be duplicated and extended as the situation demands. In this context, the new motorways and by-passes in South East London considered in Chapter 8 (8D.2) reflect the increasing need for an efficient road system. Conversely, parts of a system will fall into disuse as the demand for transport declines. This has been the case in the mining villages south of Doncaster where a network of dismantled railway lines is a feature of the landscape (Chapter 9, 9B.2).

However, in general it is possible to assess the demand for transport services by examining the density of the communications networks, noting the classes of road and number of parallel tracks on a railway.

Three different approaches to the study of communications are illustrated in this chapter.

(i) In Section 5C.2 the main categories of road in South Devon are examined and related to the physical geography of the region.

(ii) Section 5C.3 describes in detail an individual route with regard to its gradient, relative importance and landscape through which it passes.

(iii) In Section 5C.4 simple topology (see below for the meaning of this word) is used to assess the extent to which places are efficiently connected by the road network.

5C.2 · Categories of road in South Devon

5.41 Examine the network of roads on Map 5. Do you agree with the following summary which attempts to relate the main types of road to the physical geography of the Kingsbridge region?

(i) 'A' class roads form the spine of the road network. The A381, for example, follows the plateau from Kingsbridge to Salcombe. In contrast the A379 takes advantage of an east–west depression from Kingsbridge to Torcross (8242) and then clings to the coast as it proceeds to the north. West of Kingsbridge, the A379 crosses the watershed to the Avon valley.

(ii) There are few B class roads on the map. The B3194 and B3197 provide cross-country links between the major A class roads converging on Kingsbridge. The B3196 affords an additional route to the north between the A381 and A379. West of the River Avon, the B3392 terminates at the coast. At Salcombe, the sea front route is a B class extension of the A381.

(iii) The network of minor routes can be divided into two main categories based on direction and physical setting:

(a) those aligned north–south across the region, for example that linking Cole's Cross (779468) via Chillington (7943), Loo Cross (791425) and Ford (788406) with East Prawle (7836), which follows a plateau route, descending only when necessary to cross east–west orientated valleys.

(b) minor roads, less than 4 m wide, favouring the tops of ridges and linking A and B class roads with the west coast. For example, in 4 km the road between Churchstow (7145) and Bantham (6643) passes through an area almost devoid of settlement.

(iv) Service roads link isolated farms and hamlets to the road network. The lanes are often unfenced, narrow and unmetalled (grid squares 7837/7937 – 7838/7938; 7845/8145 – 7847/8147; 6845/7045 – 6847/7047).

5C.3 · Route description: general guidelines

The route taken by a road or railway line between points can be described with respect to three main features:

(i) its direction and gradient;

(ii) its relative importance e.g. category of road (A, B or unclassified), number of railway tracks (single or multiple);

(iii) the landscape through which it passes.

5.42 Study the following description of the road linking the Y-junction at Frogmore (775426) with the coast at 755405 and then say if this description is adequate. If not, what suggestions can you offer for its improvement?

This unclassified, but metalled, road is at least 4 m wide and leads south west from the village of Frogmore. It soon rises, at first slowly, but then more steeply to approximately 50 m at 772419 (note the symbol denoting a gradient of between 1 in 5 and 1 in 7 at 773422). Half way up the steeper section of the hill the road bends to the south east and then rises more slowly to the top of an east–west trending ridge at approximately 85 metres. Two hundred metres to the south west a wind pump is just visible at 776414. To either side of the fenced road there are open fields. Small woods are visible on the horizon, half a kilometre to the north east, and to the south west. This road affords extensive views over the Frogmore Creek to the north west.

At 778414 the road begins to descend and crosses a col between the headwaters of two small streams which flow respectively to the east and west. Neither valley has steep sides or a steep long-profile.

From a junction two hundred metres beyond, an unmade road leads west to the small hamlet of North Pool. At 779413 (spot height 68 m) two narrow metalled roads lead away to the north east and south east.

Beyond this col the road climbs to a Y-junction at 778409. The right fork leads west along the top of the ridge at an altitude of about 85 metres. For a kilometre and a half the road is straight and loses only fifteen metres in height. A narrow unmetalled road leads south to the village of South Pool, 500 metres away, and its church tower is visible at 776404. Three hundred metres beyond the junction, a wood flanks the northern side of the road for a distance of 400 metres.

At 765408 the metalled road swings sharply to the south. The chosen route, however, continues as an unmetalled road. Its descent to the coast at 755405, over a distance of 1.2 km, has an average gradient of 1 in 25. During this descent, the road bends southward around the head of a north facing valley and passes Halwell Farm (763407) which is protected from the north by a small wood. This route gives extensive views over the Kingsbridge estuary. At the coast the small inlet is bounded on either side by woodland. An exposed stretch of mud flats covers the floor of this inlet at low tide.

5.43 Describe the route and gradient of the minor road which leaves the A381 at 733394, follows the coast north, and rejoins the A381 immediately west of the bus station (735441) in Kingsbridge. In what respects does this route differ from that taken by the A381 between these points?

5.44 Refer to Map 1. Describe the route taken by the railway between Keymer Station (305155) and Brighton Station (310050). Use a sketch map to show the relationship of this line to the physical geography of the area through which it passes.

5C.4 · Topology and the analysis of route networks

Simplified diagrams can help us to understand the structure of route networks. These can be analysed to measure the degree of connection between a place and others in the network. This approach to the study of route networks is known as topology. It is concerned solely with the geometrical lay-out of the route system and does not consider the physical and human factors accounting for each route. Consider the following cases where the road system has been reproduced as straight lines.

Single road. Examine the A381 which links Kingsbridge with Salcombe. This is reproduced in Fig. 5.8. In Fig. 5.9 the road is represented as a straight line. Note that the settlements still appear in the same order, but the distance separating them and the direction of the road have been modified. Using the language of topology, the settlements K, W, M and S are the *vertices*, and the sections of road K–W, W–M, M–S are the *edges*.

Network of roads. Figure 5.10 shows the network of minor roads near Start Point. This

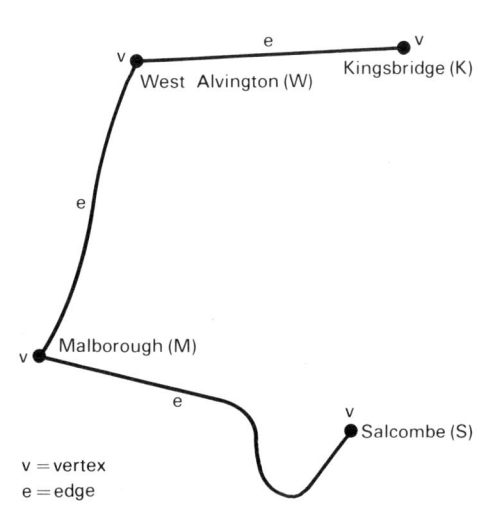

Fig. 5.8 A single road

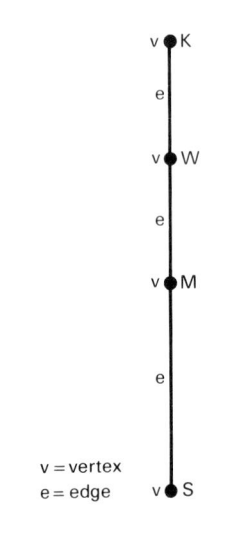

Fig. 5.9 A topological representation of Fig. 5.8

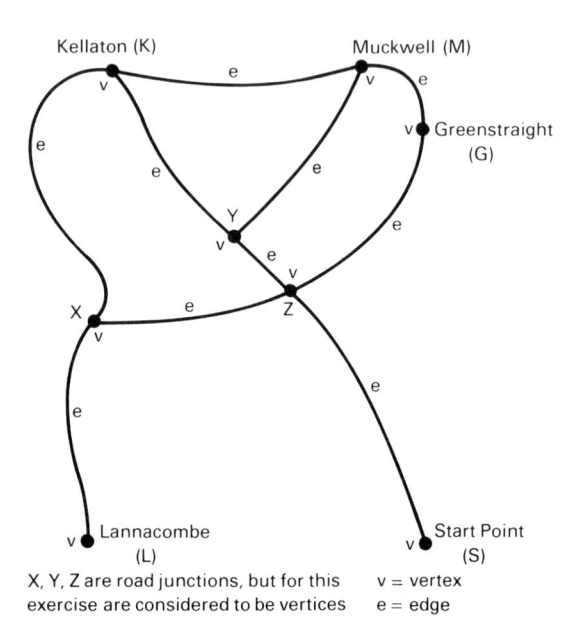

Fig. 5.10 A network of roads

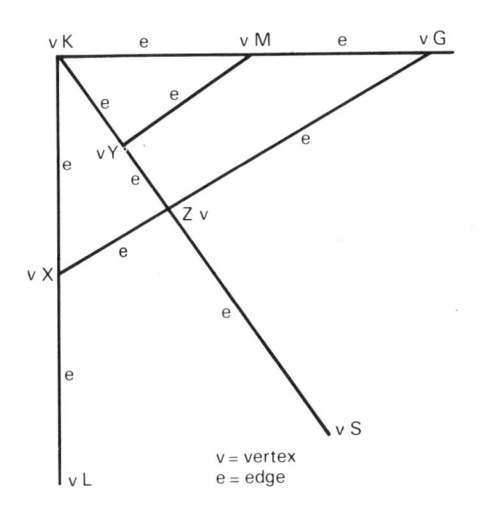

Fig. 5.11 A connected graph

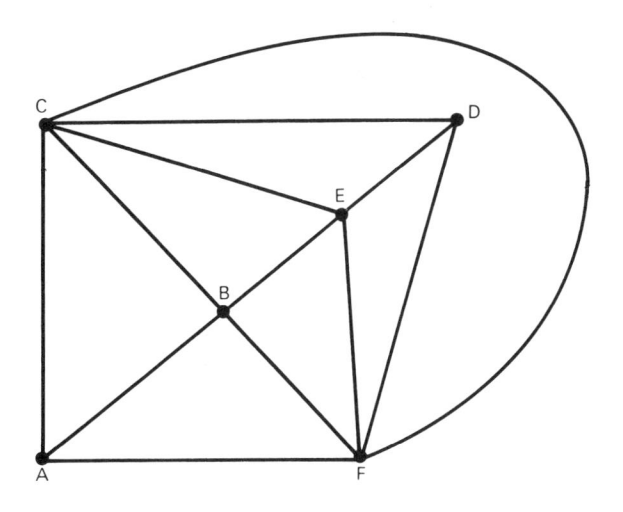

Fig. 5.12 A complete graph

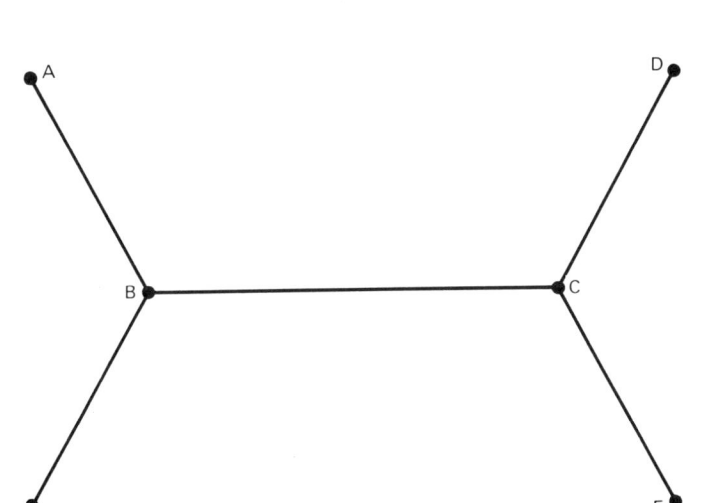

Fig. 5.13 A tree: a connected graph with no circuits

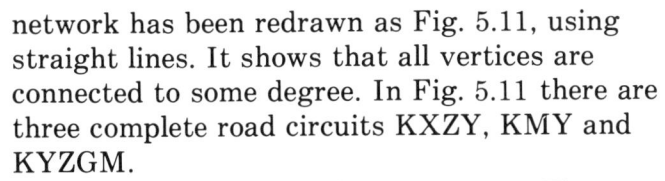

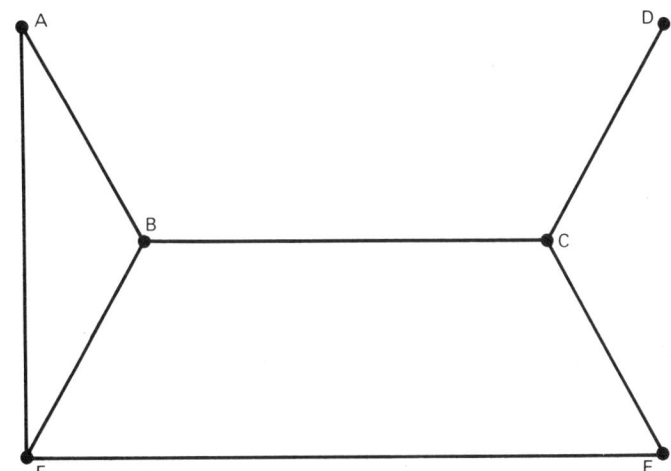

Fig. 5.14 A connected graph with two circuits

network has been redrawn as Fig. 5.11, using straight lines. It shows that all vertices are connected to some degree. In Fig. 5.11 there are three complete road circuits KXZY, KMY and KYZGM.

Now compare Figs. 5.12 and 5.13. In Fig. 5.12 all the vertices are directly connected with each other. This gives a *complete graph*. In Fig. 5.13 all the vertices are connected, but there are no circuits. Figure 5.13 is therefore known as a *tree*. In this case the addition of one more edge would produce a connected graph with one *circuit*.

There are two basic rules of topology.

Rule 1: The minimum number of edges (e) required to connect all the vertices (v) is one less than the number of vertices, or $v-1$. In Fig. 5.13, for example, the minimum number of edges (e) required to connect ABCDEF is 5 (i.e. 6−1).

Rule 2: The maximum number of edges (e) to make a complete graph (i.e. the number of edges required to connect every vertex to every other vertex) is 3 ($v-2$). Test Figs. 5.11 and 5.12 to verify this statement.

Connectivity. To measure the extent to which a transport network is connected, we can compare the number of circuits present with the number of circuits that there would be if each vertex was connected to every other vertex to form a complete graph. This relationship is summarised by the *Alpha Index* which is calculated by substituting the appropriate values for edges and vertices in the following equation:

$$\text{Alpha Index } (\alpha) = \frac{\text{number of circuits present}}{\text{maximum number of circuits}} \times 100$$
$$= \frac{e - v + 1}{2v - 5} \times 100$$

where e = number of edges; v = number of vertices

Thus in the case of the network shown in Fig. 5.14 the degree of connectivity is:

$$\alpha = \frac{7 - 6 + 1}{12 - 5} \times 100 = \frac{2}{7} \times 100 = 28.6\%.$$

These ideas on the geometry of road networks can be applied to the South Devon map to answer the following questions.

5.45 How well connected by main roads are the settlements of Kingsbridge, Malborough, Salcombe, West Alvington, Churchstow, Bridge End and Stokenham, and the point where the A381 leaves the area shown on Map 5?

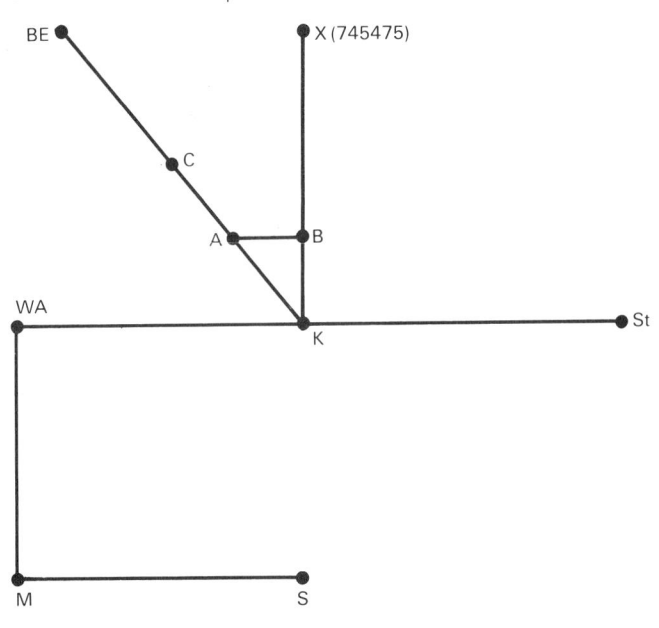

A B road junctions
BE Bridge End
     the point at which the
X A381 leaves the map
C Churchstow
K Kingsbridge
M Malborough
S Salcombe
St Stokenham
WA West Alvington

Fig. 5.15 Connectivity of A class roads in South Devon

Figure 5.15 is a topological graph of the main roads serving these settlements. The connectivity of this road network can be calculated using the Alpha Index.

$$\alpha = \frac{e - v + 1}{2v - 5} \times 100$$
$$= \frac{10 - 10 + 1}{20 - 15} \times 100$$
$$= 20\%.$$

Thus the stated network is only 20% connected.

5.46 What improvements are made to the connectivity of this road network when the B3194 and B3197 are added?

Figure 5.16 shows the revised topological graph. In

——— A class road
- - - - - B class road

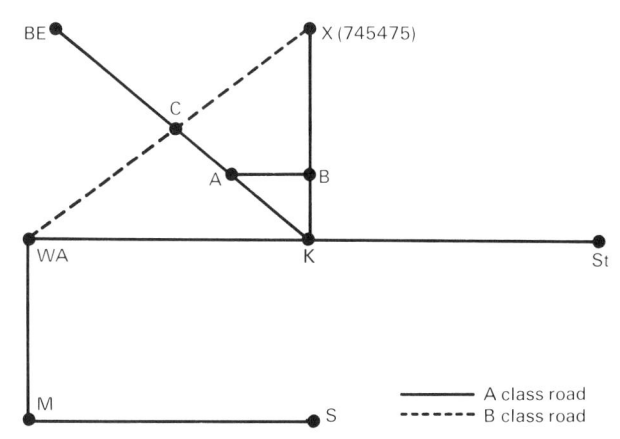

Fig. 5.16 Connectivity of A and B class roads in South Devon (for key to places, see Fig. 5.15)

this case, the connectivity is calculated using a revised value for edges:

$$\alpha = \frac{e - v + 1}{2v - 5} \times 100$$
$$= \frac{12 - 10 + 1}{20 - 15} \times 100$$
$$= 60\%.$$

Thus the stated network is 60% connected, and its degree of connectivity has increased three times.

5.47 What further improvements in connectivity might be achieved by building a bridge at Salcombe, and by widening the unclassified road which runs through Chivelstone and Kellaton to Stokenham?

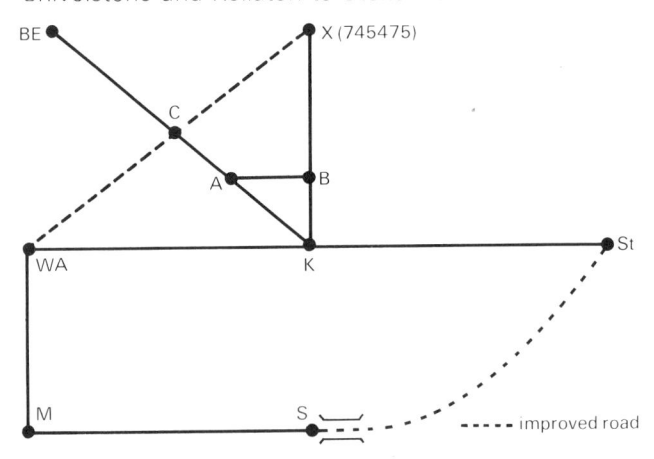

- - - - - improved road

Fig. 5.17 The final network (for key to places, see Fig. 5.15)

Figure 5.17 shows the revised network which includes a road bridge across the Kingsbridge Estuary and the proposed widening of the unclassified road. The connectivity of this network now becomes:

$$\alpha = \frac{e - v + 1}{2v - 5} \times 100$$
$$= \frac{13 - 10 + 1}{20 - 15} \times 100$$
$$= 80\%.$$

Hence, the addition of the bridge again improves the connectivity of the road network to 80%.

5.48 Network exercises of this kind can help planners to solve problems on the priority given to road improvement schemes. Imagine that you have to prepare a bus timetable and route plan for this region in South Devon. What other kinds of problem would you have to solve?

Accessibility. It is also important to find out how accessible each place (vertex) is to every other in the road network. This relationship can be shown in a shortest path matrix, such as Fig. 5.18. Here, you will see that a minimum of two *edges* separate West Allington from Stokenchurch; similarly, a minimum of two *edges* separate West Allington from point X; and so on. The number of edges separating each vertex from all other vertices is then totalled. This matrix provides the basis for calculating two indices of accessibility.

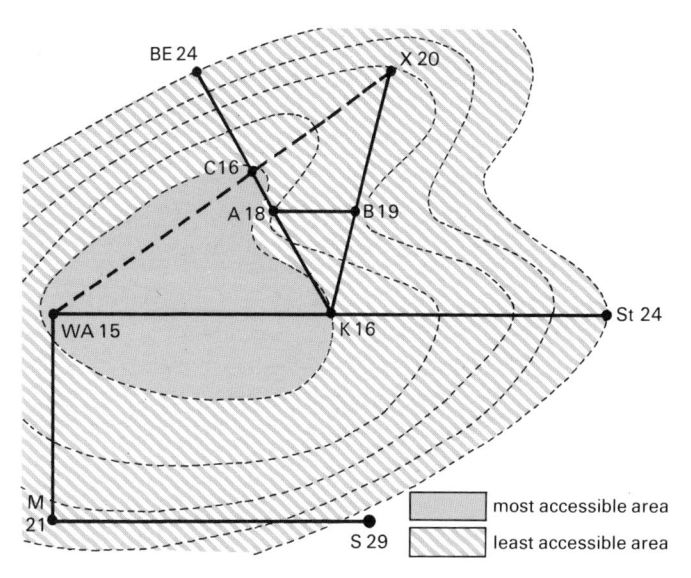

Fig. 5.19 An isopleth map to indicate the pattern of road accessibility in South Devon (for key to places, see Fig. 5.15)

	BE	C	A	B	X	WA	K	St	M	S	Shimbel Index Konig Number	Rank	Associated Number (AN)
BE		1	2	3	2	2	3	4	3	4	24	8	4
C	1		1	2	1	1	2	3	2	3	16	2	3
A	2	1		1	2	2	1	2	3	4	18	4	4
B	3	2	1		1	2	1	2	3	4	19	5	4
X	2	1	2	1		2	2	3	3	4	20	6	4
WA	2	1	2	2	2		1	2	1	2	15	1	2
K	3	2	1	1	2	1		1	2	3	16	2	3
St	4	3	2	2	3	2	1		3	4	24	8	4
M	3	2	3	3	3	1	2	3		1	21	7	3
S	4	3	4	4	4	2	3	4	1		29	10	4

Dispersion Index = 202 (total)

Fig. 5.18 Shortest path matrix (for key to places, see Fig. 5.15)

(i) *Shimbel Index or Konig Number (SI).* This index is calculated by adding together all the shortest paths from each vertex to all the others. In Fig. 5.18 it is the sum of each row.

The sum of all the Shimbel Indices is called the *Dispersion Index.* It is a guide for comparing the compactness of several networks. The lower the number, the more compact is the network.

(ii) *Associated Number (AN).* This is the shortest path, as measured by the number of edges, between a vertex and that most distant from it. It therefore gives a guide to the total compactness of the network. For example, for Fig. 5.16, the AN for Kingsbridge is 3; that for Stokenham, 4.

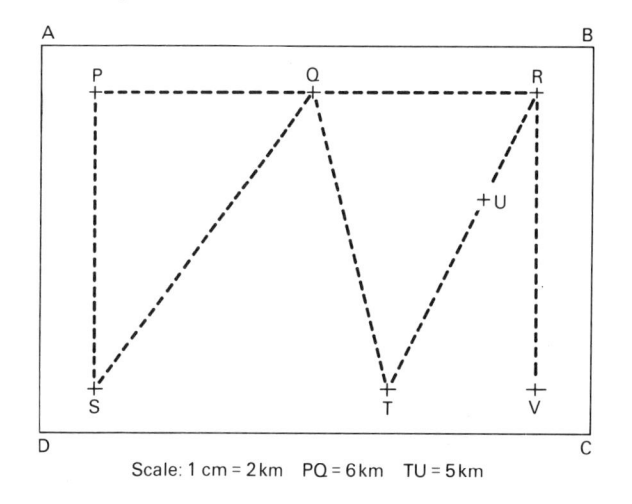

Fig. 5.20 Diagram for question 5.49

The regional pattern of accessibility can be demonstrated by preparing maps based on these index values. Here, the index values are positioned at each vertex. Then lines of equal value (*isopleths*) are drawn to isolate areas of equal accessibility. This technique is illustrated in Fig. 5.19, using Shimbel Indices.

5.49 Study Fig. 5.20 and answer the following questions:
  (i) What is the total area enclosed by ABCD?
  (ii) Give a representative fraction for the scale shown on the map.
  (iii) Which of the villages P, Q, R, S, T, U, V is the most accessible in the network? (Use the Shimbel Index.)
  (iv) How would new roads connecting (a) U to V and (b) U to V and T to V affect the answer given to (iii) above?
  (v) How well connected is the network?

Route efficiency. The efficiency of any single route can be measured by a *Detour Index*. This is derived by comparing the straightline distance separating two points with the actual route distance. Consider, for example, the A381 between Kingsbridge and Salcombe. For this route,

$$\text{Detour Index} = \frac{\text{length of A381}}{\text{straight line distance}} \times 100$$
$$= \frac{9 \text{ km}}{5 \text{ km}} \times 100$$
$$= 180.$$

This implies that the A381 is almost twice as long as it might be in connecting these centres. A Detour Index value of 100 would signify that the route was direct. (See also Chapter 6, Section 6C.1, and questions 6.29 and 6.30).

5.50 Study Map 8 between the junction of the M25, A20 (T) and the B2173 at 526677 and the junction of the A20 (T) and the B2173 at 495699 and answer the following.
 (i) In terms of the Detour Index, which of the A20 (T) and B2173 is the more efficient connecting route?
 (ii) Why then does commuter traffic use the A20 (T)?
 (iii) When and why might a motorist use the B2173?

5.51 Refer to Map 2B (Stour Valley). Draw a topological map of the metalled roads in the area bounded by 800130, 870130, 800080 and 870080. Assume that a grocer wishes to serve the villages in this area economically with a small mobile shop. From which village would you advise him to operate his business?

Density of network. Sometimes we need to compare the density of the route network in one area with that of another. This can be done by calculating a ratio between the total length of the transport network and the area it covers. Thus, for the network lying to the east of Easting 69, considered in Fig. 5.15, the calculation becomes:

$$\text{Network Density} = \frac{\text{area in km}^2}{\text{total length of network in km}}$$
$$= \frac{139.0}{40.9}$$
$$= 3.40 \text{ km/km}^2.$$

These exercises show that topology can help us to understand the efficiency of a road system and the relationship between individual places. It is just as important, however, to consider the network in terms of the quality of the roads, their gradients and the distances separating places. In this sense, map evidence can help us by suggesting those physical and human factors which have been significant in the evolution of that network.

CHAPTER 6
Central Bath

Chapter Plan

6A · **Physical setting**

6B · **Historical growth of Bath**
    6B.1 Role of historical maps
    6B.2 Physical growth
    6B.3 Features of the urban landscape (townscape)
    4B.4 Characteristics of the Central Business District (CBD)

6C · **Urban environment**
    6C.1 Access to services
    6C.2 Residential environments

6A · Physical setting

Larger-scale Ordnance Survey maps provide a great deal of information on land use in built-up areas. Sometimes, however, this detail tends to obscure the underlying pattern of contour lines. When this happens it is helpful to make a tracing of the contour pattern and drainage system. This enables us to appreciate the relationship between the built-up area and its physical setting. Figure 6.1 does this for central Bath and shows the main features of the Avon valley.

Turn to Chapter 2 and refresh your memory on the main landforms associated with the work of rivers. Now refer to Map 6 and Fig. 6.1 and complete the following tasks.

6.1 Describe the course of the River Avon between 753657 and 737651.

6.2 Examine Fig. 6.2, a line section drawn between points A and D. Note the main differences in the slopes between A and C, and C and D.

6.3 Can you explain the origin of the relatively steep gradients found immediately west of the River Avon between Eastings 750 and 753, and at Beechen Cliff (749641)?

6.4 'Dole' is an ancient word used for areas which were liable to flood. Hence the word 'Dolemead' signifies a meadow prone to flooding. Do you think this description applies to the area of land between the Avon and the 23 m contour in grid square 7564? If so, why do you think that this area might have been flooded in past times?

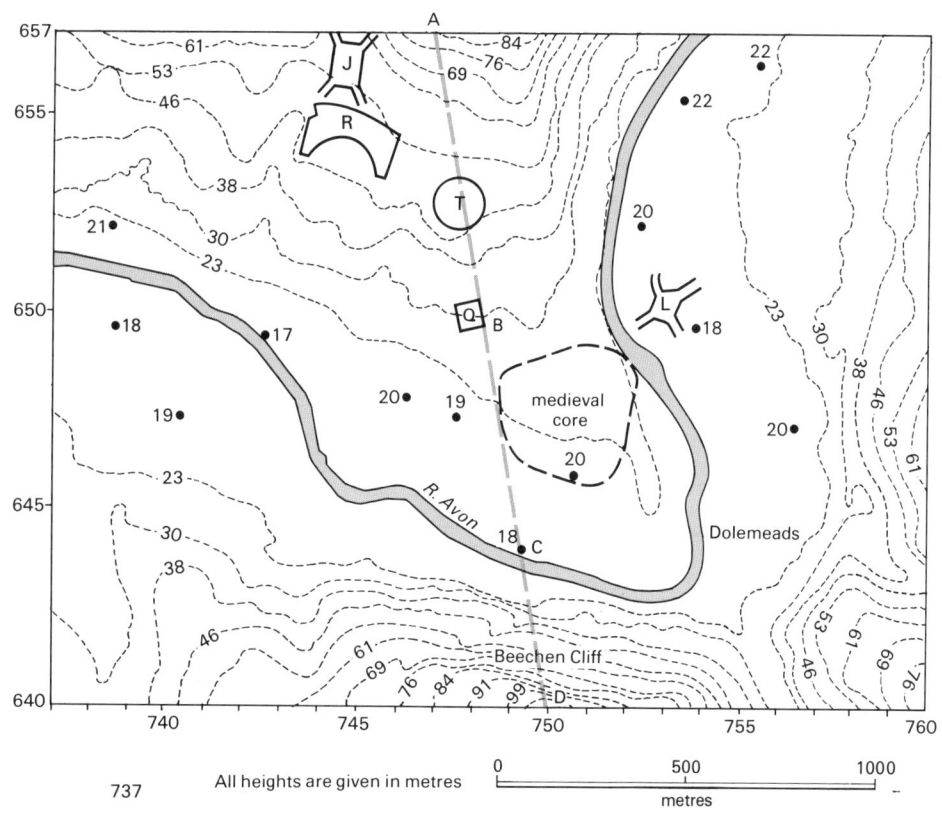

Fig. 6.1 Physical site

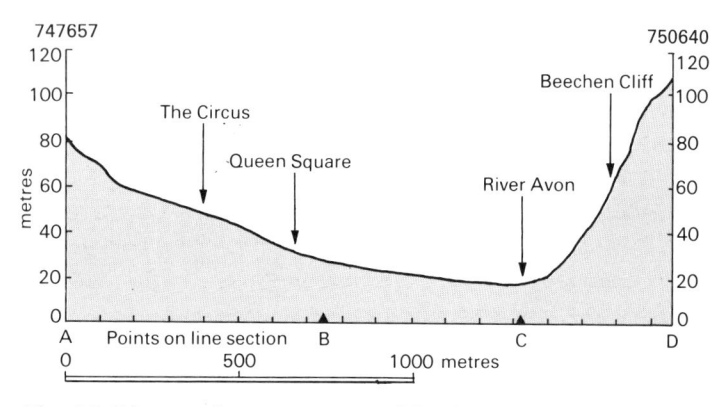

Fig. 6.2 Line section across central Bath

In general the costs of building on steep land are far greater than those incurred on level or gently sloping sites. Similarly, low-lying areas adjacent to rivers may need protection from serious flooding. With these kinds of problems in mind, it is helpful to calculate representative gradients for the built-up area in central Bath.

6.5 Use the formula given on pages 11-12 to determine the gradients between the following pairs of points: A–B; B–C; C–D; A–C. Follow the procedure below.
1. Locate each pair of points on Fig. 6.1 and the map.
2. Find the value of the nearest contour line or spot height to each point.
3. Measure the horizontal distance (in metres) between the points on the map.
4. Enter these values in the formula and calculate the gradient. Consider these results in the light of your answer to exercise 6.2.

6.6 Refer to Figures 6.1, 6.2 and 6.3 and answer these questions on central Bath.
 (i) Describe the site of the medieval core of the city.
 (ii) Offer an explanation for the elongated shape of the residential development, south of the Avon, in the period 1726–1820.
 (iii) Write a short statement to summarise the relationship between the built-up area and the physical setting of Bath in 1820.

6B · Historical growth of Bath

6B.1 · Role of historical maps

With the exception of the New Towns, most British towns and cities have developed over a long period of time. In these settlements we can therefore identify individual buildings, networks

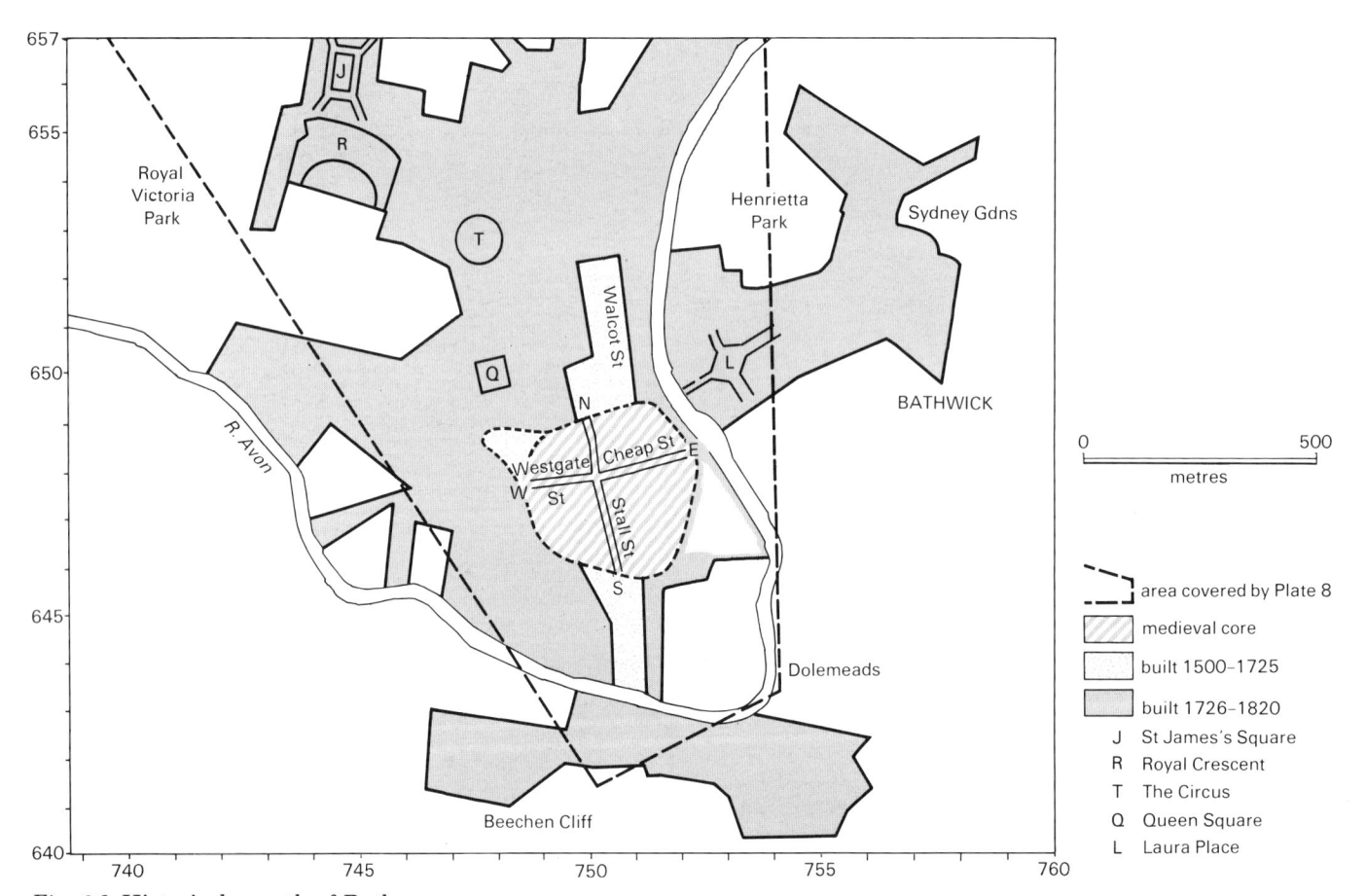

Fig. 6.3 Historical growth of Bath

of streets and styles of architecture which relate to earlier times. However, the scale and design of a map still set limits to the amount of information which is represented.

Historical maps and documents can help us to relate the stages in the growth of a town to its physical site. Such records can also give the main reasons for the principal stages in the development of central Bath. These are discussed below and the factors responsible for the expansion of the city are considered. This allows us to understand the character of the urban landscape.

6B.2 · Physical growth

The medieval town walls enclosed the Roman settlement. Westgate Street–Cheap Street and Union Street–Stall Street represent the main Roman streets which linked the principal gates of the town. By 1725 a minor ribbon of building had extended north from the town wall along Walcot Street. A second area of building proceeded from the South Gate to the important bridging point on the Avon. Far more extensive residential growth occurred between 1725 and 1820. Notice, too, that an area of parkland seems to have prevented large-scale building in the north west. South of the river there was only limited development between the bridging point and Beechen Cliff. Bathwick was designed as a suburb on the Pulteney estate. Work on this scheme commenced in 1788.

Figure 6.3 shows the area of Bath covered by Plate 6. Repeat exercises 6.7 and 6.8 using first the map and then the air photograph. For guidance, refer also to Fig. 6.2 and Fig. 6.3.

Plate 6

6.7 Identify Union Street, Stall Street, Westgate Street and Cheap Street.

6.8 Attempt to follow the line of the medieval walls in the present-day network of streets.

6.9 Refer to the map extract. Describe the route followed by the railway track between 737645 and 759657. What additional information on the route of the railway line can be gained from the air photograph?

6.10 Examine the site of the main railway station (753644). Why do you think the station was located at this point?

6.11 Refer to Plate 6. Contrast the structures of the railway bridges at map references 751643 and 754644.

6B.3 · Features of the urban landscape (townscape)

This account covers the period from the Roman occupation to the middle of the 19th century. In this period many of the distinctive features of the townscape in central Bath were first set out.

Springs of hot mineral water are found near the city centre in Bath. These have been very important in the development of the city as a spa. They were first exploited by the Romans who established a health resort, *Aquae Sulis*, in the 1st century AD. This settlement was later enclosed by the medieval walls. In AD 973 Edgar was crowned King of England in Bath Abbey (751648). On several occasions in the 16th and 17th centuries English royalty visited the spa. In their wake came large numbers of titled and wealthy visitors. Many had houses built in the city for their summer-time use.

For the benefit of these leisured visitors the city authorities provided assembly rooms, pump rooms, parks, gardens, promenades, libraries, theatres and many other facilities. The distinctive townscape of Bath evolved in this period. Famous architects, including John Wood and Thomas Baldwin, were responsible for the design of many public buildings and private houses. They adopted the Classical architectural styles of ancient Greece and Rome. The builders engaged for this work used a soft, light-coloured, oolitic limestone which was quarried nearby.

In the period 1720–1870 the road communications between Bath and other parts of Britain were greatly improved. Canal and railway links were also introduced. These strengthened the attraction of Bath as a spa and had an important effect on the economy of the city. The River Avon was canalised as far as Bristol in 1727 while the completion of the Kennet–Avon Canal (757644) in 1810 connected Bath by water with London. Bath was linked into the Great Western Railway system in 1841; a connection with the London, Midland and Scottish Railway came later in 1869. The presence of these railways encouraged the location of a range of industries on the floor of the Avon valley.

As a result of these changes, by the mid 19th century Bath had become an important commercial centre. Large numbers of shops, offices and other services were clustered inside the medieval walls. The city served an extensive area, or hinterland, in the county of Avon. In addition it supported a large number of private schools. Specialised hospitals were also established. These used spa water for the treatment of rheumatic and other complaints.

6.12 Look at Plate 6 and Fig. 6.3. Name the Georgian residential developments J, R, T, Q, L. Give their map references.

6.13 How many storeys have the houses in Laura Place?

6.14 Describe the site of the Royal Crescent using the guidelines on page 105.

6.15 Prestigious town houses of the Georgian period often had their own stables and coachhouses ('mews'). These were usually sited at the end of the garden. Does the map show evidence of 'mews' developments in (i) The Royal Crescent (ii) The Circus (iii) Queen Square (iv) Great Pultney Street, east of Laura Place?

Suggest the purposes for which these mews buildings are used at present.

6.16 Before the introduction of legislation to control air pollution, the soft oolitic building stone in central Bath was readily blackened by smoke. Examine the map and air photograph. What possible sources of air pollution might be suggested from the evidence of land-use patterns? How might the physical geography of the valley have aggravated this problem? What weather conditions would be most likely to cause a build-up of atmospheric pollution in the valley bottom?

6.17 Examine the valley floor to the west of Easting 75. How can you identify industrial land uses in this area? Draw a boundary for this industrial region. List the problems you faced in doing this exercise.

6.18 Prepare an overlay on tracing paper to show the distribution of schools and hospitals. Comment on these patterns. Is it possible to identify different kinds of school from the evidence on the map and air photograph?

6B.4 · Characteristics of the Central Business District

A concentration of large shops, important offices and public buildings occurs in the central parts of most British cities. Urban geographers have named this area the Central Business District (CBD). (See Fig. 7.4.) It is a densely built-up region in which are found some of the highest buildings in the city. This region benefits from its accessibility to most other parts of the city. Bus routes tend to converge on the area. For those using private transport, car parks are normally available in places adjacent to the CBD. Moreover, the main railway station may be found at the edge of this area. Although shops are not identified on the map, many public buildings and offices are indicated in those parts of Bath built-up before 1820 (Fig. 6.3).

6.19 Examine the area situated between the River Avon and Northing 65, and to the east of Easting 745. List all the named public buildings in this area. Compare this list with a similar list prepared for the area between Northings 650 and 657, and Easting 745 and the River Avon. What are the main differences between the two lists?

6.20 Compare the map and air photograph. Give the map references for, and identify the use of, the buildings marked A, B, C, D and E.

6.21 Determine the location of the car parks on Plate 6. Mark these on a tracing overlay of the map. Comment on the pattern. To what extent are public buildings served by nearby car parks?

As a town grows in importance and size, its CBD will expand (Chapter 7 Section 7C.1). The CBD, however, rarely grows in neat, concentric rings. Sometimes it will extend along the important roads which converge on the city centre. On occasions, too, an old residential area near the CBD may be comprehensively redeveloped for shops and offices. This process has recently occurred in Bath between the South Gate 751645 and bridges spanning the Avon.

It is unfortunate that the map cannot give detailed information on these changes in land use. Nevertheless, we can see from the air photograph where new buildings have been erected. In some cases, too, we can determine the actual use of an individual building. The newer buildings are generally lighter in colour. Many have flat roofs. These features are in contrast with the darker stonework and ridged roofs of the older buildings elsewhere in the city. The tower blocks of flats and offices on the western margin of the area indicate further changes in land use. It would seem, moreover, that some plots of land are being used as ground-level car parks, possibly on a temporary basis until needed for redevelopment.

6.22 With the help of Fig. 6.1, describe the topography of the area between the South Gate (751645) and the River Avon.

6.23 Examine the shape, relative size and frontage of the building marked B on Plate 6. What is this building? Give evidence to support your decision.

6.24 Do you think that the proximity of features B and C, and D and E on Plate 6 reflect good planning practice? Justify your answer.

6.25 Select evidence from the air photograph to demonstrate that measures have been taken to reduce the risk of floods in central Bath.

6C · Urban environment

6C.1 · Access to services

From time to time the families living in central Bath will need to use local facilities, including shops, a post office, school or church. It is very inconvenient for many, especially mothers with young children and the elderly, if these services are located far from the home.

Let us consider, for example, the distribution of post offices in residential areas. Prepare a tracing overlay of the map to show the location of all post offices. Draw a circle, radius 0.25 km, around each post office, including the Head Post Office (750649). Examine this overlay, and answer question 6.26.

6.26 Which major residential areas on the map are more than 0.25 km from a post office?

We know from our experience that a straight line distance of 0.25 km will invariably underestimate the actual distance a person will need to walk to visit a post office. Consider the case of an elderly person living at the southern end of Dorset Street, near the railway line (737645). The straight line distance between this house and the Post Office on the Lower Bristol Road (739647) is 330 metres. In comparison, the shortest road distance along Dorset Street and the Lower Bristol Road is 400 metres. The relationship between the straight line distance separating the pensioner's home and the post office, and the shortest road distance, can be summarised by calculating a *Detour Index* (See page 66). In our example:

$$\text{Detour Index} = \frac{\text{shortest route distance}}{\text{straight line distance}} \times 100$$
$$= \frac{400}{330} \times 100$$
$$= 121.$$

This indicates that the pensioner would have to walk approximately 1¼ times the direct distance to visit the post office.

6.27 What would a Detour Index of 200 signify?

6.28 What would a Detour Index of 100 signify?

6.29 Now consider the following journeys and answer the related questions:

(i) Living next door to the Post Office in the Lower Bristol Road (739647) is a student who cycles to College (748646) each weekday. Calculate the Detour Index for this journey.

(ii) At the junction of Archway Street and Broadway (756645) lives a person who wishes to visit the Police Station (753645). He decides to drive his car by the shortest possible route. He parks in the car park immediately north of the Police Station (see Plate 6). Calculate the Detour Index for this visit.

(iii) Each day a person living at the junction of Edward Street (755651) and Great Pultney Street walks to the Library (751649). Calculate the Detour Index for this journey.

What effect does the River Avon have on the journeys made by these three imaginary persons?

6.30 Does the railway line act as a barrier between the city centre and the residential areas of Oldfield Park (738641) and Westmoreland (743644)? Support your judgement by analysing representative journeys from houses in each area to specified public buildings in the centre of Bath.

6C.2 · Residential environments

We may all hold personal views on the desirability of any area as a place in which to live. The map, however, suggests interesting contrasts in the various local environments of central Bath. These may be examined using indicators of environmental quality. Study the list of indicators in Table 6.1 and consider the meaning of each one. Discuss these with a friend, and then answer questions 6.31 to 6.34.

Table 6.1 Indicators of environmental quality

House type:	Style of house – detached, semi-detached, terraced, etc.
	Size and position of gardens.
House site/ situation:	Altitude.
	View and aspect.
Local environment:	Street gradients in the immediate locality.
	Density of residential/non-residential buildings.
	Through roads or cul-de-sac situation.
	Width of local roads.
	Nature of traffic flows e.g. main roads; roads leading to industrial areas.
	Proximity to industrial areas.
	Presence of trees and access to public open space e.g. parks, gardens, recreation grounds.
	Noise pollution e.g. railways, main roads.
	Access to services and facilities e.g. post offices, schools, churches.

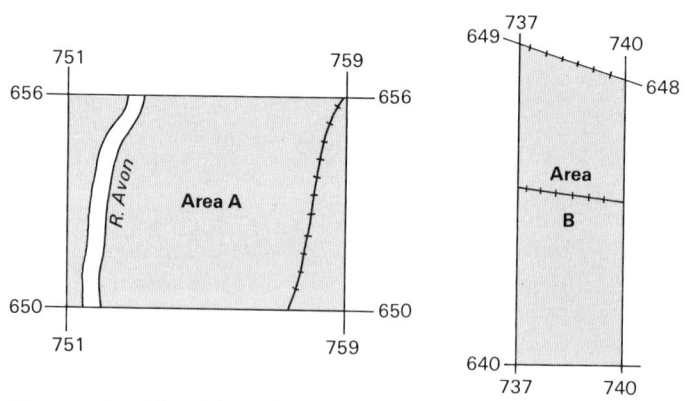

Fig. 6.4 Residential environments: test areas

6.31 Identify on the map the areas marked as A and B on Fig. 6.4. Compare and contrast the street patterns in these localities.

6.32 Can you add other indicators of environmental quality to the given list? If so, explain how these can be measured from the map.

6.33 Which of the following areas offers the better residential environment for an elderly person: Dorset Street (737646) or Park Lane (737654)? Justify your answer.

6.34 Describe the residential environment of a person living in Henrietta Gardens (755653).

CHAPTER 7
Lower Teesside

Chapter Plan

7A · **Physical background**
  7A.1 Height of the land surface
  7A.2 Drainage system
  7A.3 Land reclamation

7B · **Industrial land use**
  7B.1 Identification of industrial units
  7B.2 Industrial regions and communication networks
  7B.3 Selection of an industrial site

7C · **The process of urban growth**
  7C.1 Land-use regions in British towns
  7C.2 Settlement patterns on Teesside

7A · Physical background

7A.1 · Height of the land surface

In the lower valley of the Tees (Map 7), the distributions of settlement and communications tend to conceal the contours and network of drainage. On closer examination, however, it can be seen that while the height of the land varies little over much of the map, it rises steeply to over 200 m in the south east. This contrast is clearly shown on Figures 7.1 and 7.2, a contour trace and representative line section (marked X–Y on Fig. 7.1), respectively.

7.1 Are these two methods effective in representing the pattern of relief on lower Teesside? Refer to Chapter 1, Section 1B.3, and justify your answer.

Values for spot heights and triangulation pillars have been included on Fig. 7.1 to complement the contours. It can be seen that approximately two-thirds of the region lies below the 20 m contour while almost half lies below 10 m. There is a high water table in lower Teesside. This is confirmed by areas of surface water and several named tracts of marshland (4619; 4924; 4925). Figure 7.2 demonstrates the pronounced break of slope between the floodplain and coastal marshes, and the higher region in the south east. It also highlights the width and flatness of the broad floodplain on the right bank of the Tees.

7.2 What do you observe about the location of spot heights with regard to: (i) their distribution over the entire map, (ii) actual sites.

7.3 Draw an annotated line section between 453255 (30 m) and 480187 (12 m) Comment on the problems you encountered in constructing this diagram.

7A.2 · Drainage system

Main characteristics. In general, to the north of the Tees and below the 10 m contour the drainage flows sluggishly to the east. Artificial drainage channels, shown as straight blue lines, and embankments are distinctive features in this landscape (5124; 5125; 5025). In addition there are several bodies of standing water. Some of these may supply water to large industrial units (5223; 5023).

Billingham Beck, a left-bank tributary, drains the area west of Easting 49. It flows through a relatively shallow valley and has several minor tributaries, some of which have been artificially channelled or embanked. The lower section of the Beck has been straightened, possibly to allow navigation by small vessels or to protect the large industrial plant on its left bank from flooding.

South of the Tees, a number of near parallel water courses drain the higher land. Characteristically, these streams have steeper gradients. In this region the process of urban expansion has considerably modified the drainage system.

Drainage in the urban environment

7.4 What has been the impact of urban and industrial development on the natural drainage system?

To answer this question, refer to Fig. 7.1. Locate the streams labelled A–D, and for each, describe its course to the River Tees, using the following guidelines.

1. Examine changes in the cross profile and gradient of the valley.
2. Seek evidence of: artificial straightening of the channel; modifications to the banks (e.g. embankments); diversions in direction (e.g. to avoid railway lines, major roads, etc.); diversions underground (if possible, give the distances for which streams are underground).
3. Prepare a short written statement on each stream, giving evidence of human modification to

the drainage system. Indicate the form that such modifications have taken. Illustrate your answer with appropriate sketch maps and give map references to features of particular interest described in your report.

Now use the guidelines above to answer questions 7.5 and 7.6.

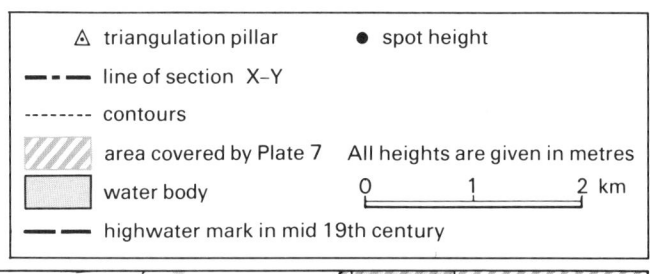

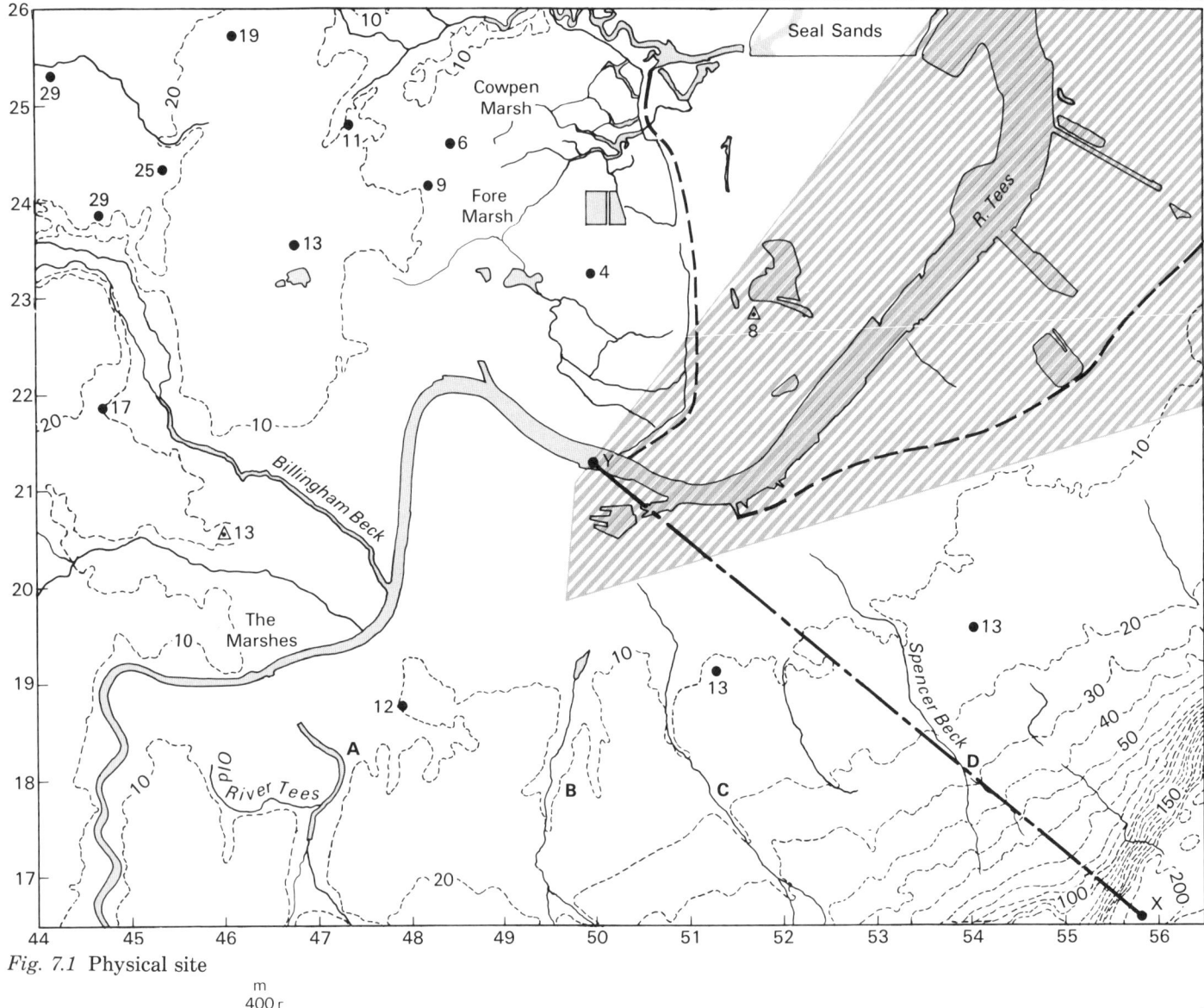

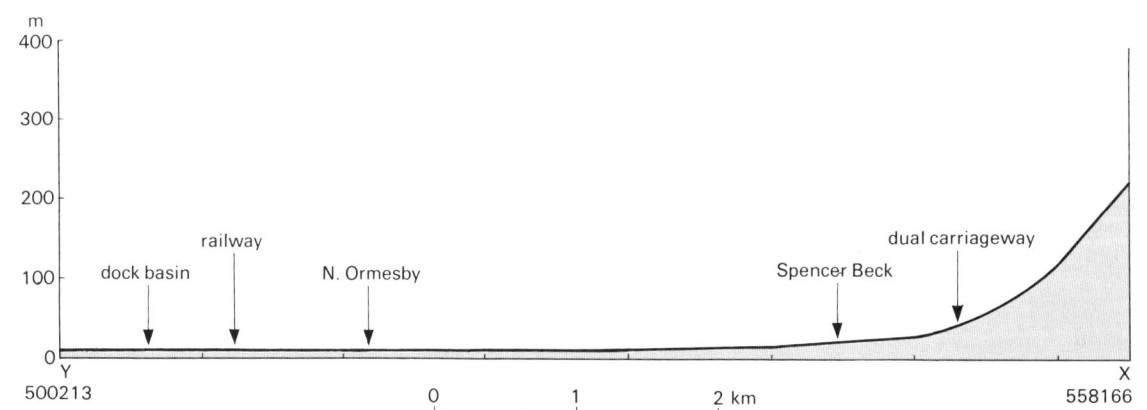

Fig. 7.1 Physical site

Fig. 7.2 Line section 500213–558166

7.5 Refer to Chapter 8, Section 8B.2 and explain how the growth of South East London has affected the pattern of drainage. Draw sketch maps to illustrate your answer.

7.6 Examine the water courses between the following points on Map 9 (South Yorkshire): 535936 – 608937; 556955 – 598975. Write a short account of each stream. Indicate the extent to which its course has been modified by human activities. Illustrate your answer with sketch maps.

Flood protection. Floodplain environments have posed many problems for human settlement throughout history. Early settlements were normally built on dry-point sites to minimise the loss of life and reduce the damage caused by flooding to houses and land. During the Industrial Revolution many British towns grew rapidly. In several instances the floodplains of large river valleys were covered with housing and industry. Precautions were therefore needed to safeguard these investments. The measures adopted included:

(i) the construction of embankments and retaining walls to keep the river in its normal channel;

(ii) dredging and deepening of the river bed, especially where water transport was important;

(iii) making cuts through the necks of meander loops, in the lower course of a river, to increase the flow of floodwater and to assist navigation.

These measures were especially important in the tidal reaches of a river, where high tides could retard the flow of water and lead to flooding.

7.7 Bear these comments in mind and examine, carefully, both banks of the River Tees between 447165 and 545260. Now undertake the following exercises.
   (i) Prepare an overlay on tracing paper of both river banks.
   (ii) Mark the sections of each bank which have been reinforced by artificial embankments.
   (iii) Indicate the areas where the banks have been straightened. (Refer to Plate 7 to help with this exercise).
   (iv) Identify the sections of each bank which form part of an industrial site.
   (v) Mark all dock basins, jetties, loading stages, and wharfs. Explain how you have identified these features. What help did you get from the air photograph?
   (vi) Trace the drainage pattern between the River Tees, A19(T) and Northing 17. Examine, in detail, the course of the Old River Tees. Can you offer an explanation for this physical feature?
   (vii) Why are jetties needed on the left bank of the River Tees down river from Easting 50, but not on the right bank?

7A.3 · Land reclamation

Large areas of coastal marshland fringing the principal estuaries in Britain have been reclaimed for industrial purposes. What evidence is there of this process in the lower Tees valley?

The area south of Seal Sands (5225) displays many characteristics of a reclaimed estuarine area. It has:

(i) a straight and seemingly unnatural northern edge which extends for almost 2 km (517255 – 535255);

(ii) dykes to protect its western flank from the sluggish eastward flowing drainage of Cowpen Marsh and Foremarsh (e.g. 5124);

(iii) a generally low-lying surface;

(iv) an absence of natural surface drainage.

This physical region extends south to the River Tees and supports several very large industrial plants: two oil refineries (5223, 5225), chemical works (5324) and a tar distillery (5121).

Downstream from Easting 52, on the right bank of the Tees, the land surface is equally low-lying. There is no evidence on the map to indicate that this land has been reclaimed from the sea. However, the position of the high water mark in the mid 19th century, shown on historical maps, confirms that this area has been reclaimed from coastal marsh (Fig. 7.1).

7.8 Relate the high water mark of the mid 19th century shown on Fig. 7.1 to the map extract. Can you find evidence in the drainage pattern to indicate the previous position of the coastline? Was the position of the coastline in the 19th century an important factor which determined the routes chosen for major roads and railway lines?

7B · Industrial land use

7B.1 · Identification of industrial units

Not all industrial land uses are shown on an Ordnance Survey map. Some industrial units, of course, are labelled. It is difficult, however, to study the industrial composition of an area where there are no named industries or where the printed information refers only to a 'works'

(548167, 480258) or 'industrial estate' (4723). In these circumstances we must obviously use our common sense, experience of industrial land uses, and knowledge of regional geography to interpret the available evidence. For example, areas of larger-scale extractive industry such as quarrying (Map 3, 6164) or mining (Map 9, 5492/5592) will leave visible scars on the landscape. Other activities like oil refining can be recognised by the symbols for chimneys and oil storage tanks arranged in an orderly pattern (Map 7, 5523). Power stations give rise to a network of overhead transmission lines (Map 8, 5977).

Thus the printed evidence on maps can provide some clues to the existence and possible identity of an industrial unit. Study the following guidelines and answer the related questions.

Guidelines for identifying industrial units

1. *Type of building and the use of surrounding space*: size of building(s) and spatial layout; use of surrounding space – storage, communication, waste disposal; surface area of site; function of particular buildings on site.

2. *Communication linkage* (including those important in the past, for example, a dismantled railway line, canals): road access; railway sidings; canal wharf; river/port; access to jetty or dock basin; overhead electricity transmission lines.

Plate 7

3. *Site and situation*: physical site and situation (nature of land surface and its location); relationship of the site to the local and regional communications network; evidence of similar industrial units nearby; kinds of named industrial units in the vicinity.

7.9 Use these guidelines to prepare a short description of each of the named industrial units on Map 7: oil refinery 5223; 5523; chemical works 5324; tar distillery 5121; industrial estate 4620.

7.10 On the north bank of the River Tees, confined between the river, A19(T) and B1275, there is a large industrial plant. Describe its spatial arrangement. What kind of industrial unit do you think it is? Give reasons for your answer. What problems have you experienced in reaching a decision?

7.11 Give reasons for the locations of the oil refineries (5523; 5223; 5225).

7.12 Identify, from the map, the principal land uses in grid square 5423.

7.13 Examine the limitations of the site occupied by the industrial unit lying between the River Tees and the main railway line in grid square 4518.

7.14 List the potential benefits enjoyed by firms located on the industrial estates in grid squares 4620 and 4723/4724.

7.15 What does LC mean (e.g. 553228; 454188)?

For the following questions use information from Plate 7 and Map 7:

7.16 Name the industrial activities A, B, C and D on Plate 7, and give their map references.

7.17 Give map references for the new industrial buildings labelled F and G on the photograph.

7.18 Contrast the industrial land-use patterns on either bank of the River Tees.

7.19 What type of industrial plant is E? Use the map evidence in grid squares 5421/5422/5622 to help with your answer.

7.20 Analyse the pattern of industrial land use in grid square 5020. Explain how the air photograph can help in identifying particular industrial land uses.

7B.2 · Industrial regions and communication networks

Figure 7.3 was prepared using the guidelines in Section 7B.1. It shows that all the major industrial units are located in areas below the 10 m contour. Downriver from the Transporter Bridge (500213), there are several large industrial plants on reclaimed land.

Fig. 7.3 Principal communications and industrial land use

An adequate network of communications is important to the success of an industrial region. Those seeking sites for new industrial units will also be interested in the range of transport modes – road, rail, air, water – and the plans proposed for their development.

As Fig. 7.3 shows, railway lines serve both banks of the River Tees. East of Cargo Fleet (5120), on the south bank, the railway route follows the old shoreline.

Four road bridges cross the lower Tees. The A19(T), a dual carriageway, is the main north–south artery of road communication. A second north–south route uses the Transporter Bridge (5021) to link the A172 and A178. South of the Tees, this single carriageway cuts through densely built-up residential and industrial areas. Unlike the A19(T), however, it lacks elevated interchanges at the major road junctions. Down-river from the Transporter Bridge, on the south bank, a rectangular grid of A class roads serves the industrial region and neighbouring residential areas (Fig. 7.3). These major roads also provide an effective link with other parts of Britain.

Since the mid 19th century, the communications network has been extended and improved to meet the needs of industry and a growing urban population. It seems that certain of the major industries were sited on reclaimed land to secure additional access to ocean transport.

East of Billingham, on the north bank, spurs have been added to the road and railway networks to service the reclaimed areas at Seal Sands and the major industrial plants on the estuary.

7.21 Search the industrial regions delimited on Fig. 7.3 and find examples where:
 (i) the main railway network has been extended by mineral and service lines;
 (ii) dock basins and jetties have been constructed;
 (iii) road networks have been improved by dual carriageways, by-passes and interchanges.

7.22 How important to the south-bank industrial region is the main railway line between 440178 and 565234? What kinds of industry are served by this line?

7.23 Examine the road and railway network to the east of Easting 50. What extensions have been made to these networks to service the industrial sites on reclaimed land?

7.24 Prepare an overlay on tracing paper to show the network of electricity transmission lines. Comment on the origin and destinations of these lines.

7.25 Compare and contrast the A19(T) between 468165 and 455260, with the A172/A178 between 512165 and 510260.
Comment on the:
 (i) type and design of the carriageway;
 (ii) alignment of the route with regard to physical features;
 (iii) possible obstructions to making a quick journey by car.

7B.3 · Selection of an industrial site

7.26 Imagine that you are the Industrial Development Officer for an international petrochemical company. You are responsible for selecting a site for a new chemical plant on lower Teesside. The site must conform to the following specification:

Isolation: for safety reasons, it must be at least 2 km from existing populated areas.
Site area: approximately 1 km², the arrangement of buildings will be such that the plant must be at least ½ km wide.
Communications: rail access is needed and also a metalled road. Some raw materials will be transported by pipeline from an existing oil refinery. Chemicals will be exported by rail and small coastal vessel.
Costs: For financial reasons, it will not be possible to build more than 2 km of new railway track, 3 km of metalled road, 1 jetty and 3 km of pipeline to an existing refinery.

Now undertake the following tasks.
 (i) Prepare a sketch map to show feasible locations for the new chemical works.
 (ii) Justify your choice of one site.
 (iii) Comment on the problems that a civil engineering contractor might experience in preparing this site and building the industrial plant.
 (iv) List the types of problems you would envisage for the successful operation of this plant.

7C · The process of urban growth

It is important to know about the ways in which towns grow when seeking to interpret the landscape in a heavily urbanised region.

7C.1 · Land-use regions in British towns

Studies have shown that there are broad similarities in the spatial patterns of land use in British towns. As Fig. 7.4 shows, the Central

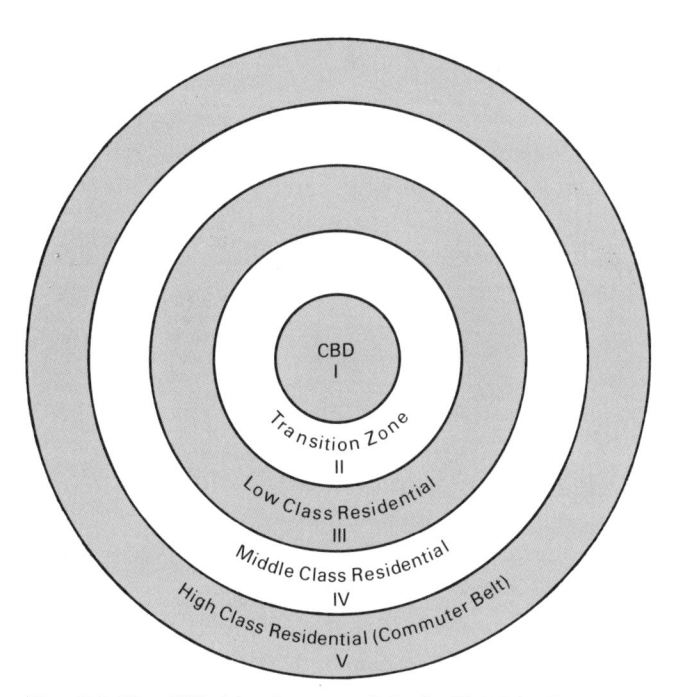

Fig. 7.4 Simplified land-use model of a British city

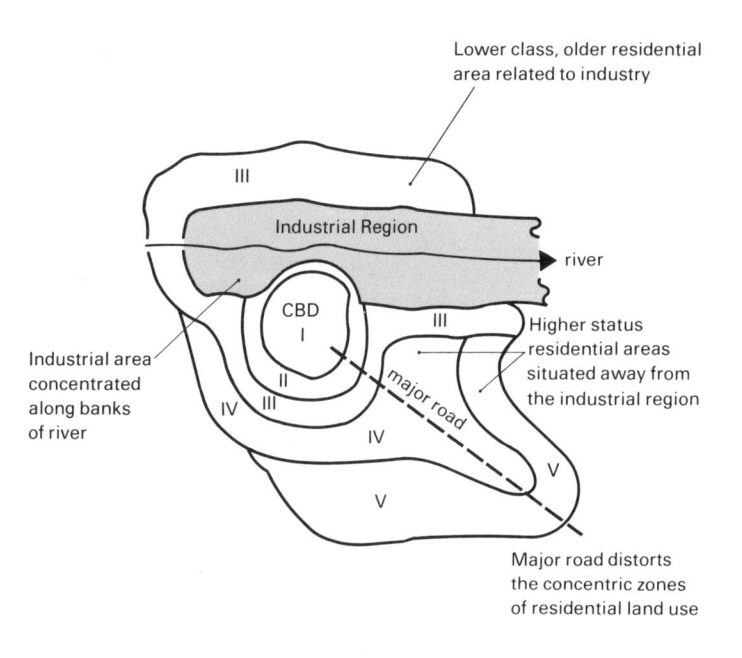

Fig. 7.5 Modified model of land-use regions in a British city (see Fig. 7.4 for key to labels I–V)

Business District (CBD) is the most accessible region in the town. In this area are concentrated important public buildings, offices, large department stores and facilities for commercial entertainment. It is normally served by a bus station, railway station and car parks. Usually surrounding the CBD is a belt of wholesaling establishments and small industrial units. Next there is a girdle of higher-density terraced housing. This area, in turn, is surrounded by newer areas of lower-density development. Finally, around the edge of the built-up area are found the homes of commuters to urban employment.

Figure 7.4 is, of course, a very generalised land-use model. For several reasons, individual towns may have different spatial arrangements of land use. These include:

(i) the influence of the physical site which can distort the width and concentric arrangement of these zones;

(ii) very specialised land-use regions in some towns e.g. port areas, university quarters, heavy industrial zones, etc;

(iii) planning legislation and practice which can influence the arrangement of land use in some areas;

(iv) New towns which have been planned from the beginning will be different from long established towns.

This land-use model can nevertheless help us to understand the main features of urban land-use patterns and their spatial arrangement (see Fig. 7.5).

7.27 Prepare a simplified land-use map for
(a) Brighton/Hove (Map 1)
(b) Middlesbrough/Billingham (Map 7)
Show the CBD, industrial regions and residential areas.
   Compare these maps and answer the following questions:
 (i) What are the main similarities and contrasts in the types of land use in these urban areas?
 (ii) What factors in the sites of each settlement account for the shape of the land-use regions?
(iii) Why are they different in some ways from the simplified land-use model given in Fig. 7.4?

7.28 Write a short account of the land-use patterns within 0.5 km of either side of Billingham Beck. Give reasons to explain the absence of large-scale industrial, residential and recreational activity on the valley floor.

Refer to Plate 7 and answer these questions:

7.29 Give map references for the buildings H and J.

7.30 Comment on the nature of the land use at K.

7.31 Identify a vehicle park and give its map reference.

7.32 What evidence is there of new commercial buildings (e.g. offices, shops). Support your answer with map references.

7.33 Refer to Map 1, and locate the area shown on Plate 1B. Draw a simplified land-use map of this area to show
(i) major contrasts in the density of the built-up area,
(ii) residential areas of different ages.

As towns grow, there will be changes in the size and arrangement of the main land-use regions. For example, an increased population will demand additional services from the CBD. This region will expand outwards (and upwards with taller buildings) into the surrounding girdle of light industry and wholesaling establishments. In turn, this zone will spread outwards into neighbouring residential areas. Chapter 6 (6B.4) examines this process of land-use change in more detail using a 1:10 000 map of central Bath.

7.34 What changes in land use have taken place in the following locations on Map 7?

488200	493205	516193
545207	447228	498172

Although an O.S. map and air photograph may not give sufficient information to define all the zones shown in Fig. 7.4, we can normally distinguish three contrasting zones of urban land use in British towns. These include the CBD, industrial regions, and residential areas of different density and design.

The Central Business District (CBD). In many British towns the CBD covers at least part of the historical core of the settlement. We can determine the location of the CBD from a map by:

(i) the convergence of important roads, and the sites of the railway station and bus station;

(ii) an assemblage of important public buildings e.g. Town Hall, Museum, Law Courts, Library, Head Post Office.

In historical towns, moreover, the area is characterised by an irregular network of streets and cluster of churches (possibly, too, a Cathedral).

Large-scale maps provide more information on the size of the CBD and its range of services. Compare, for example, the information relating to CBD activities in Bath (Map 6, Scale 1:10 000), 745646/745650 – 754643/752650) with that for Dartford (Map 8, Scale 1:50 000, 5474), Brighton (Map 1, Scale 1:50 000, 300035/300060 – 320035/320060), Kingsbridge (Map 5, Scale 1:50 000, 7344) and Middlesbrough (Map 7, Scale 1:50 000, 4920).

Industrial regions. The distinctive land-use characteristics of industrial regions and their relationship to the network of communications have already been discussed in Section 7B.

Residential areas. To explain in detail the historical growth of residential areas in a settlement would require a local knowledge of its economic, social and political history. Nevertheless, it is possible, using map evidence, to identify the relative ages of the main residential regions in most British towns. These characteristics are summarised, diagrammatically, in Fig. 7.6.

(i) *Historical core*: In historical towns the original core of the settlement usually forms part of the CBD. This area often has an irregular network of streets with little open space. Housing is far less important than commercial and administrative land uses, although some people may live in flats above shops and offices.

(ii) *18th century and 19th century*: Significant changes occurred in the landscape of many British towns in the period of the Industrial Revolution. These changes included:

(a) the introduction of land uses related to improvements in transport e.g. additions to the road system, extension of the railway network with the construction of stations and related transport facilities;

(b) the development of new zones of industrial land use, for example factories, ports and mines;

(c) the building at high densities of suburbs of terraced housing for those working in the new industrial regions. In these areas the streets were often set out on a grid-iron pattern.

(iii) *20th century*: During the 20th century several new elements have been introduced into the British townscape.

(a) The building densities in new residential areas have generally been lowered. Semi-detached and detached housing units are now more common than terraces.

(b) Widespread suburban development has continued. This relates to the availability of public transport services and higher levels of car ownership.

(c) Curvilinear and cul-de-sac street patterns have become a common feature in peripheral housing estates. Some large estates now have their own schools, shopping centres and health-care facilities.

(d) Planning control has been important, especially since 1918, in influencing the growth of towns in certain directions. It also controls the

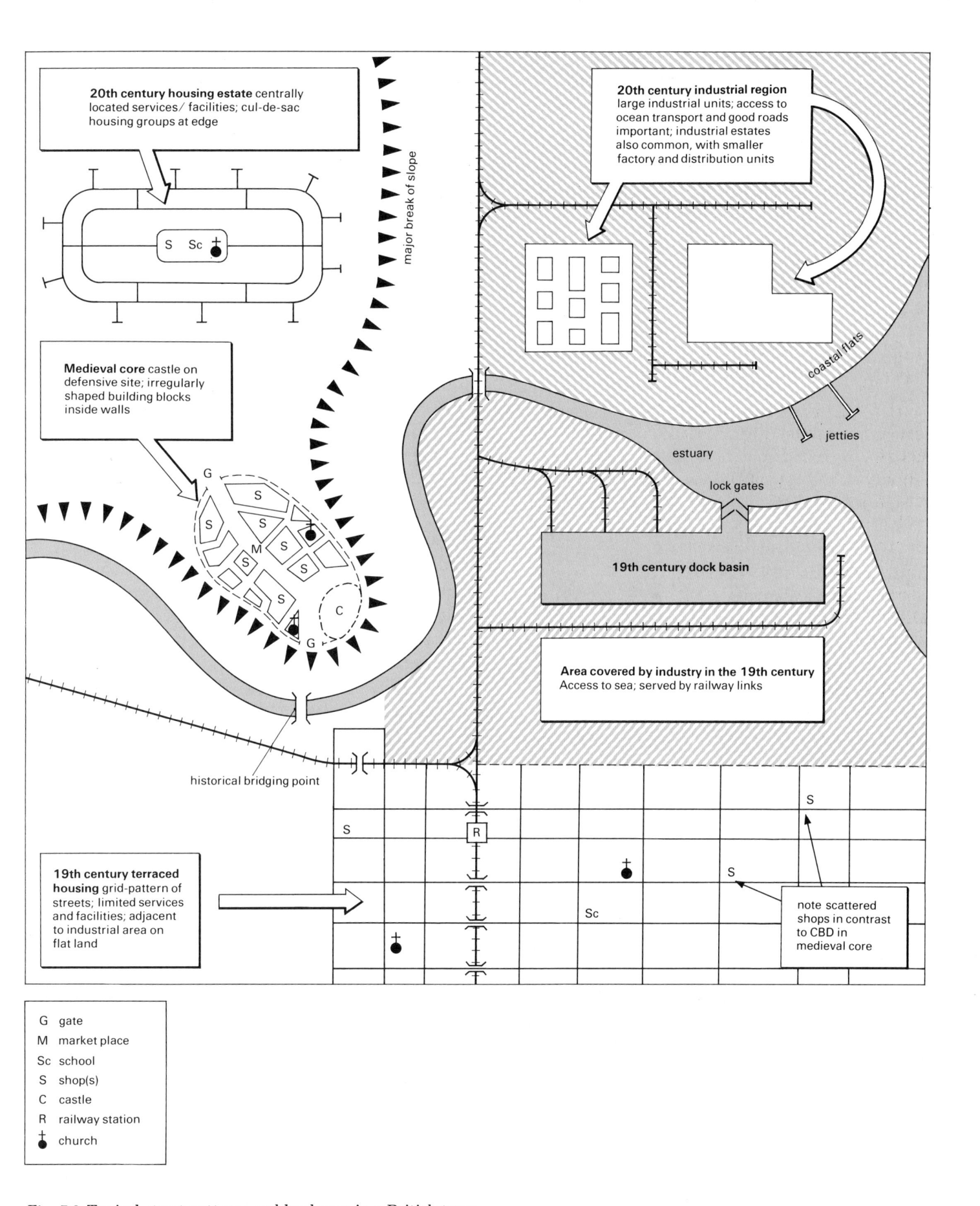

Fig. 7.6 Typical street patterns and land uses in a British town

pattern of land use within the built-up area.

(e) Nearby villages have expanded to become dormitory settlements for many working in the town.

7C.2 · Settlement patterns on Teesside

Figure 7.8 indicates that the industrial and residential areas on Teesside are well segregated. Rural land uses, however, intrude into the built-up area along the major watercourses in the south, since the danger of flooding prevents use of the area for building purposes. This happens, too, on the north bank, where the Billingham Beck breaches the riverside industrial region. Elsewhere small enclaves of rural land use exist in the built-up area: for example, between South Bank (5420) and Teesville (5419); in grid square 4820; and the park situated to the north east of Linthorpe (4919).

To appreciate the growth of settlement on lower Teesside we can use the guidelines for dating street plans summarised in Fig. 7.6 and described above. Figure 7.7 shows the spatial distribution of the grid-iron street plan. This design, a characteristic of 19th century planning, is most common to the south of the river. Here, it extends over several kilometres between Linthorpe (4818) and the river bank. There are

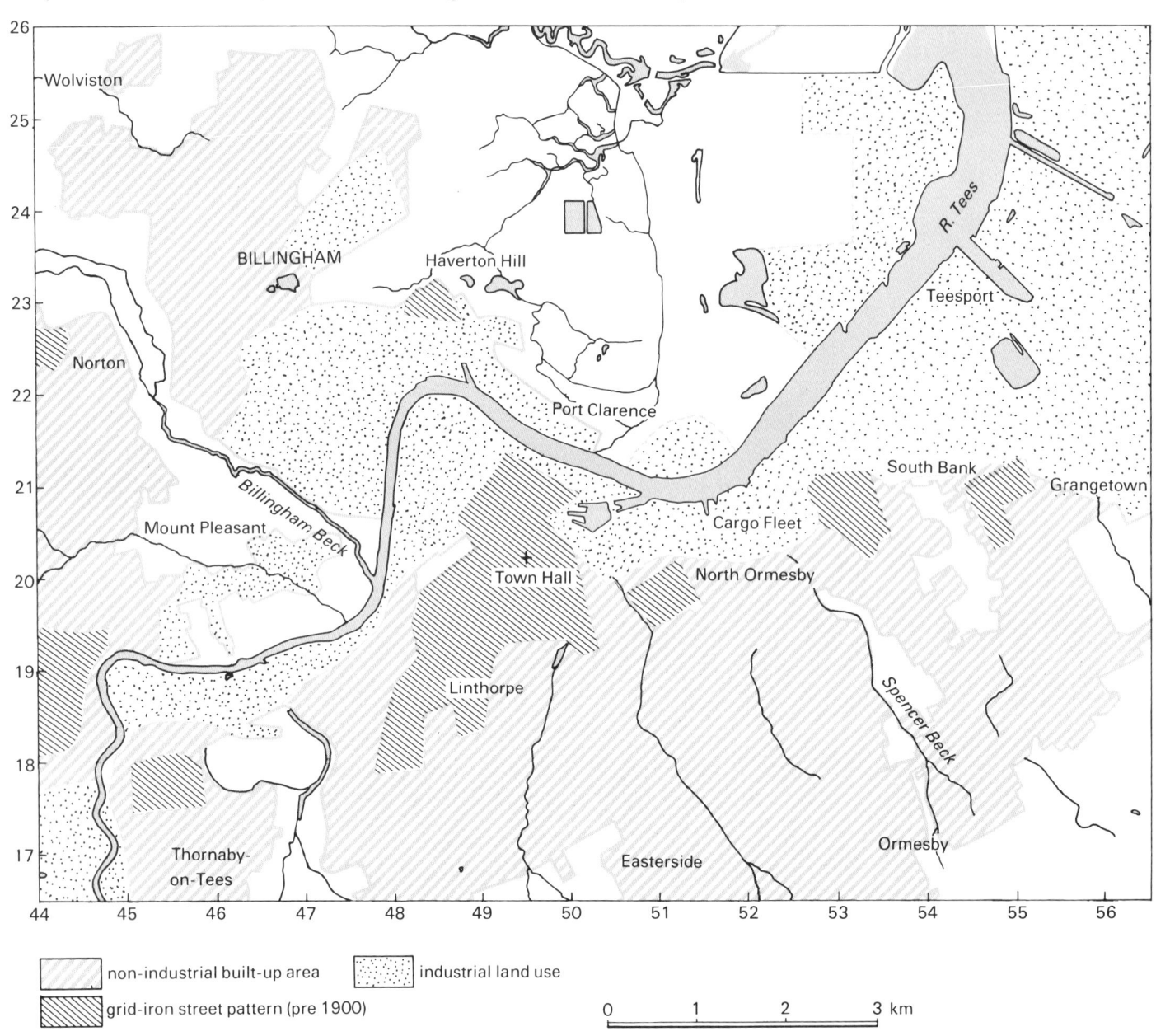

Fig. 7.7 Distribution of grid-iron street plans

substantial areas, too, at North Ormesby (5119), South Bank (5320) and Grangetown (5520). North of the Tees, there is little evidence of the grid-iron pattern of streets, except in the extreme west.

As the grid-pattern residential areas lie adjacent to the industrial regions, it seems likely that this housing was built for the families of those working in river-bank industries.

Although the size and orientation of the building blocks varies from one place to another, there is little open space included in these residential areas. The park lying north east of Linthorpe is an exception (4919). Moreover, it is interesting to note that the grid-pattern of streets immediately south west of the Transporter Bridge (500213) has been set out around a central square (495211), suggesting perhaps a planned residential area even in the 19th century.

Since 1900 residential growth has advanced across open countryside to the north and south of the Tees (Fig. 7.8). In this process, villages like Ormesby (5317), occupying more elevated sites, have become engulfed by urban sprawl. In the north, less than 0.5 km of open countryside separates Wolviston (4525) from the northern edge of Billingham. Dual carriageway by-passes skirt both of these settlements.

In the 20th century several large housing

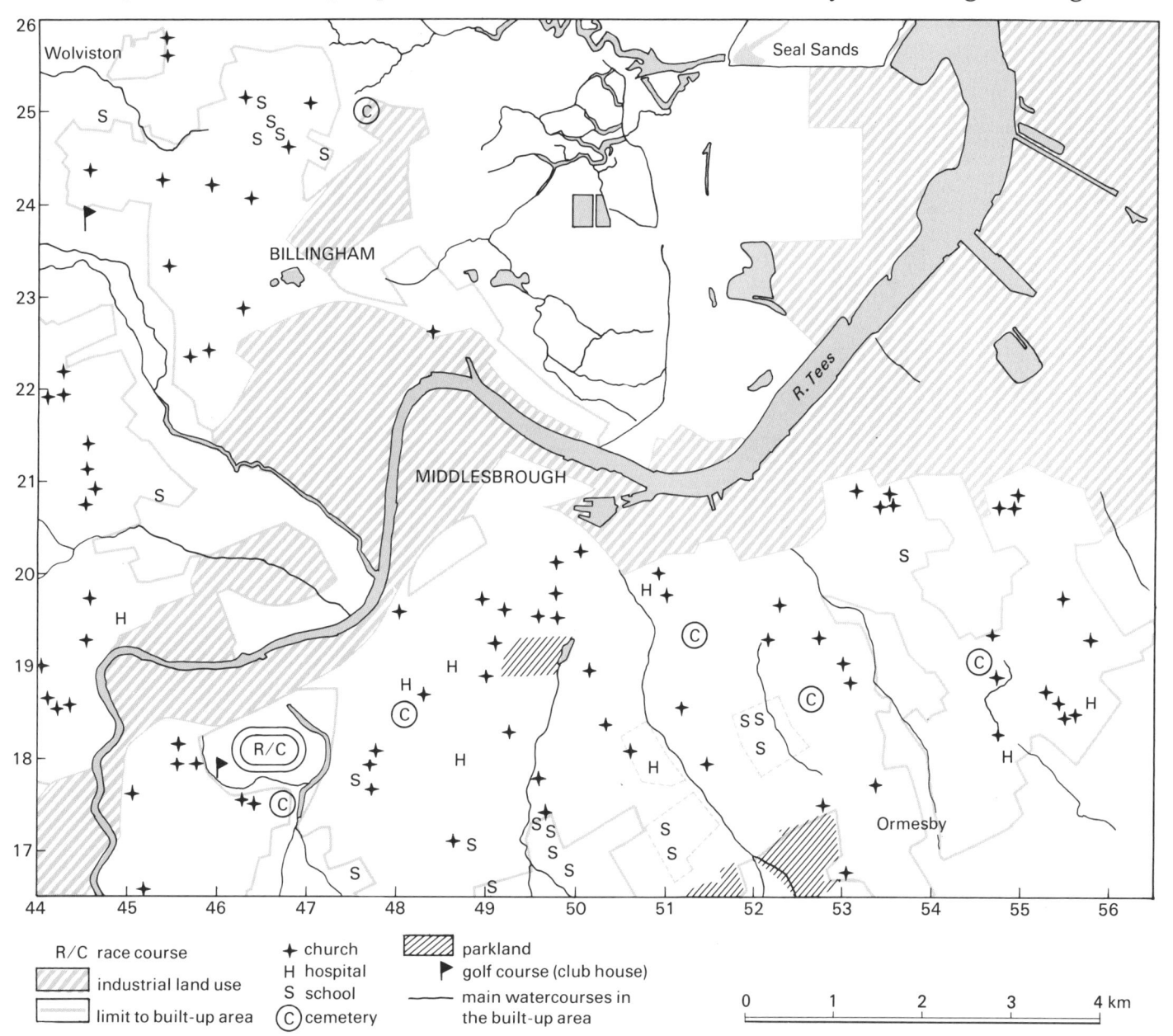

Fig. 7.8 Distribution of residential and industrial land use

estates have been constructed near the edge of the built-up areas (e.g. 4623 – 4625; 5219; 5320; 5318; 5317). These are served by schools, churches and presumably neighbourhood shopping centres. Tracts of open land in the shallow valleys have been used by services requiring large amounts of space, for example schools and hospitals. There are instances, too, where several schools share a common site (4625; 5117).

7.35 Refer to Chapter 6. Use Table 6.1 to assess the main advantages and disadvantages of living in the Easterside housing estate (500165). Repeat this exercise for Mount Pleasant (4520), Wolviston (4525) and Grangetown (5520).

7.36 Why do schools (e.g. 5117, 4624, 490165) occupy such large sites? Suggest reasons to explain why such large sites were available at the time the schools were built.

7.37 Compare and contrast the street patterns in the following grid squares on Map 7: 5520; 4919; 4624; 4424.

7.38 Describe the street plan and house types found in the eastern part of grid square 4920 on Map 7. How old are the houses? Support your answer with appropriate evidence from Plate 7.

7.39 Examine Plate 1B which covers part of Hove on Map 1.
 (i) Name the land uses A, B, C.
 (ii) Name the residential areas X and Y.
 (iii) Identify areas of land use which relate directly to the tourist industry at Hove.

CHAPTER 8
South East London

Chapter Plan

8A · **Introduction**

8B · **Density of the built-up area**
  8B.1 Density contrasts
  8B.2 Land surface and drainage pattern
  8B.3 Land use transect

8C · **Land use on the urban–rural fringe**

8D · **Road and rail communications**
  8D.1 Basic network
  8D.2 Road communications
  8D.3 Railway communications

8A · Introduction

As a consequence of the continued growth of British towns and cities, large areas of farmland have been converted to residential, industrial and recreational land uses. In addition new roads and railway tracks have swallowed up tracts of agricultural land. This chapter combines the evidence from a 1:50 000 (Second Series) map of South East London (Map 8) and a vertical air photograph (Plate 8) to examine three topics: firstly, contrasts in the density of the built-up area; secondly the effects of urban growth on rural land-use patterns; and thirdly changes in the communications network associated with urban growth.

8B · Density of the built-up area

At first sight we may find it difficult to digest the great variety of information shown on Map 8. For this reason, diagrams have been prepared to highlight important features of the built-up area:

(i) a simplified map to indicate the extent of the built-up area and enclosed open spaces;

(ii) a contour trace to show the pattern of relief and drainage;

(iii) annotated line sections to portray contrasts in the density of the built-up area.

8B.1 · Density contrasts

The proportion of the map covered by urban land uses declines from west to east. Figure 8.1 distinguishes two contrasting regions of land use.

Region I lies west of the River Cray. It is heavily built up, although in some localities like Eltham, Bromley and Chislehurst, there are substantial areas of open space and woodland. To the north, however, in Bexley and Erith there is proportionately less open space.

Region II extends east from the River Cray. Here, the built-up area is far less continuous and more fragmented. Although the main settlements of Dartford and Swanley contain little open space within their boundaries, large areas of productive agricultural land separate them from smaller settlements like Hextable, Crockenhill and Sutton-at-Hone. East of the Darent, and south of Dartford, the landscape becomes distinctly more agricultural and contains only minor settlements like South Darenth and Horton Kirby. North of Dartford the sparsely populated areas of marshland bordering the Thames (Dartford Marshes 5476; Stone Marshes 5675) provide sites for industry, including a Power Station (5676) and Works (5476).

8.1 Draw a sketch map to show the distribution of industrial land uses on both banks of the Thames, to the east of Easting 50. Mark the main lines of drainage, railways, motorways and A class roads, and the 20 m contour. Comment on the kinds of industries you have noted. To what extent are these related to the River Thames?

8.2 Name the settlements A, B, C, D and E on Plate 8. (Refer to Fig. 8.2 and Map 8.)

8.3 Refer to Table 6.1, page 72. Use these guidelines to compare the residential environments in the following grid squares: 4776; 4370; 5668; 5277.

8B.2 · Land surface and drainage pattern

A comparison of Fig. 8.1 with Fig. 8.2 shows that in general the building densities are lower on land below the 20 m contour. For example, there is relatively little building on the floor of the Cray Valley downstream from Foots Cray (4770). Crayford, however, is an exception. Much

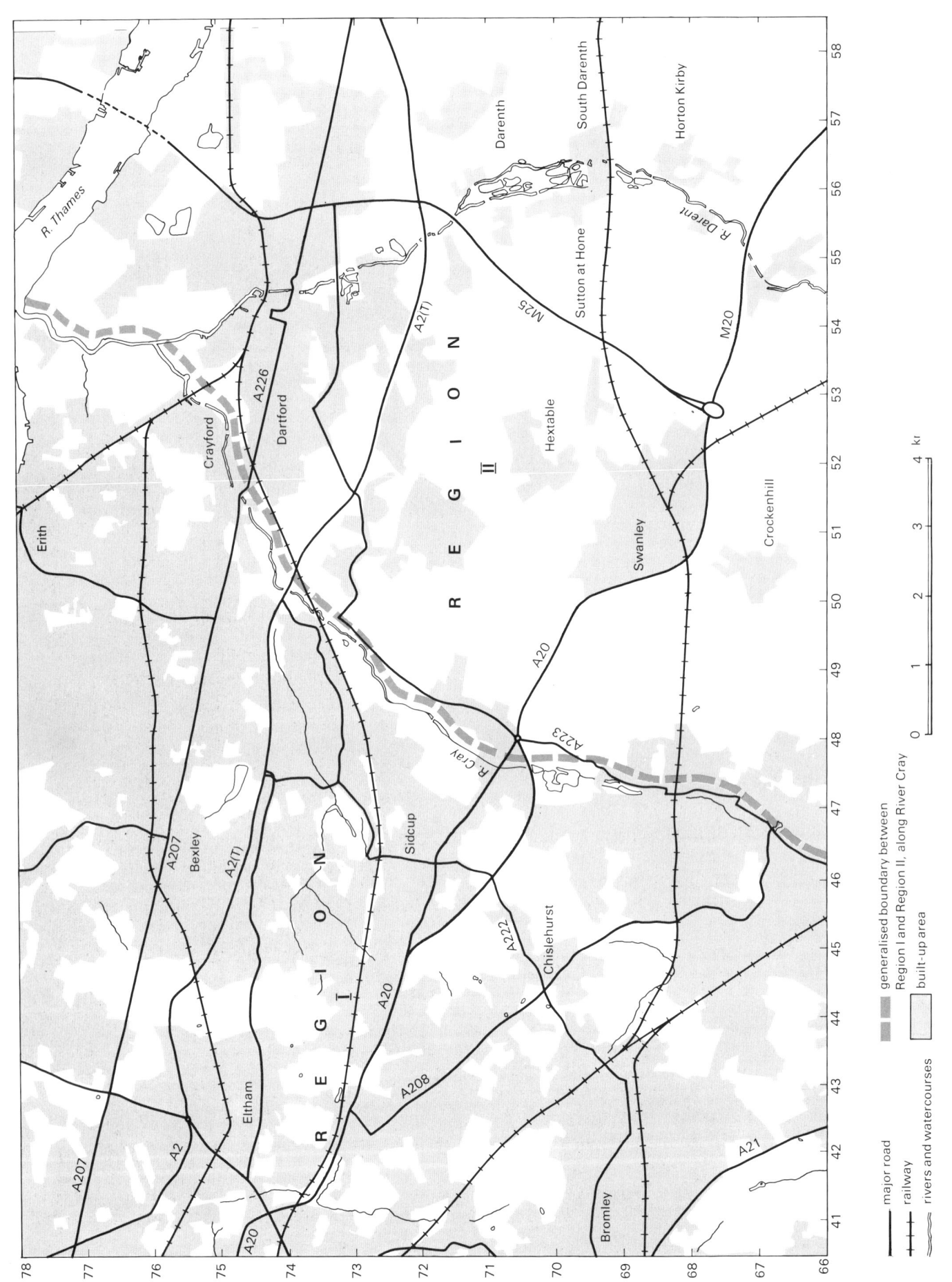

Fig. 8.1 Generalised outline of the built-up area

Map 7 Lower Teesside 1:50 000

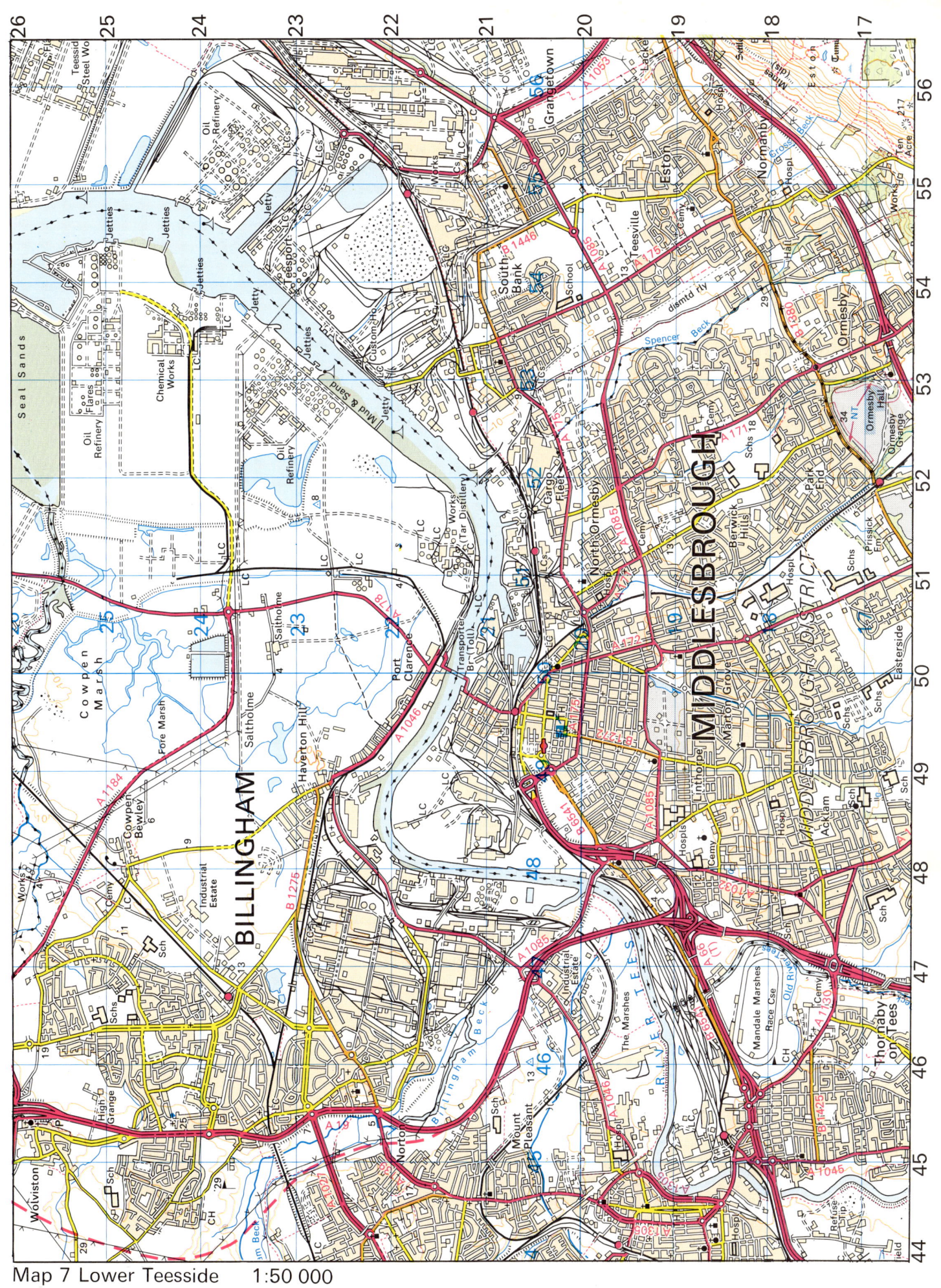

Map 8 South East London   1:50 000

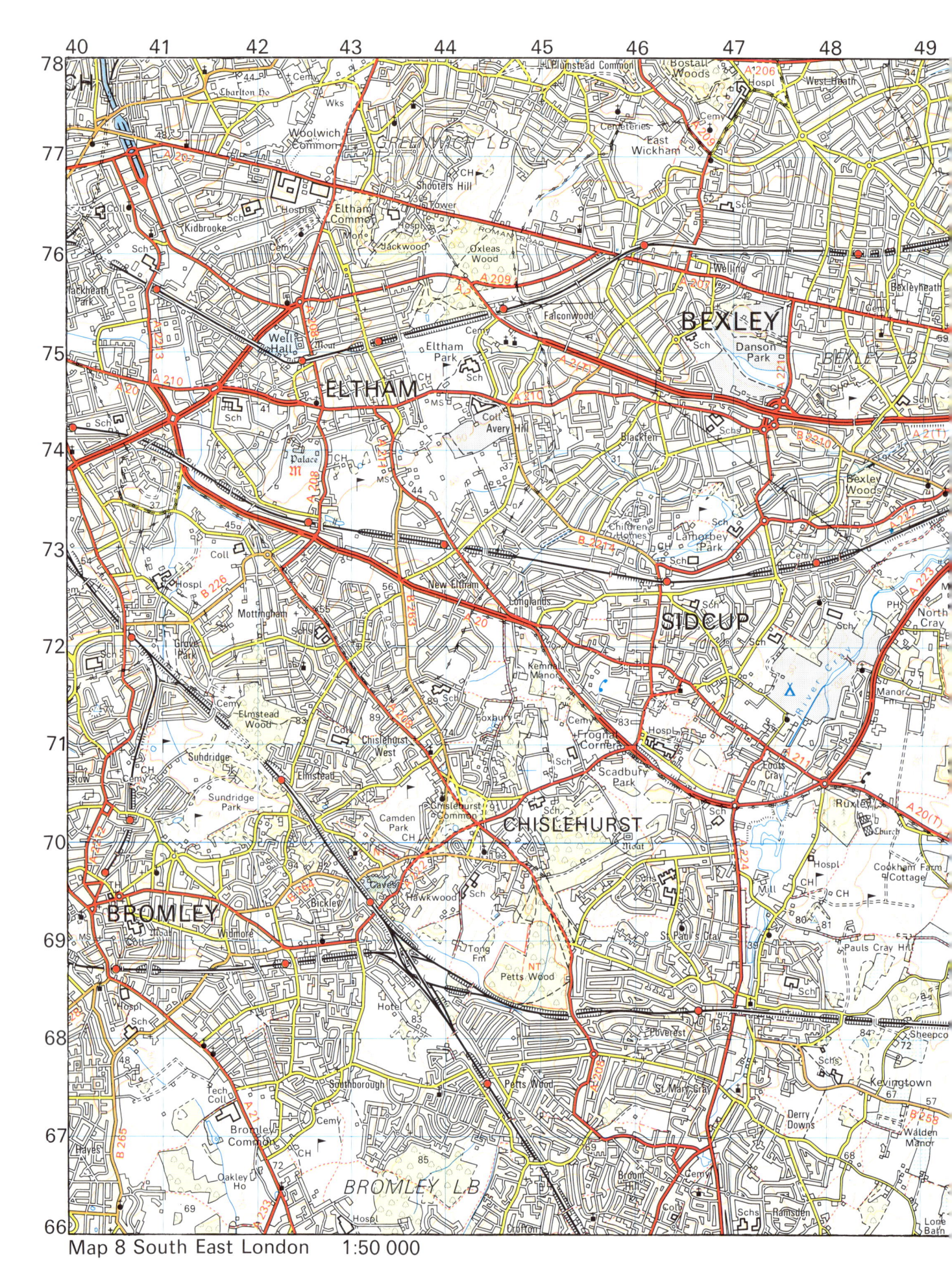

89

Map 9 South Yorkshire 1:50 000

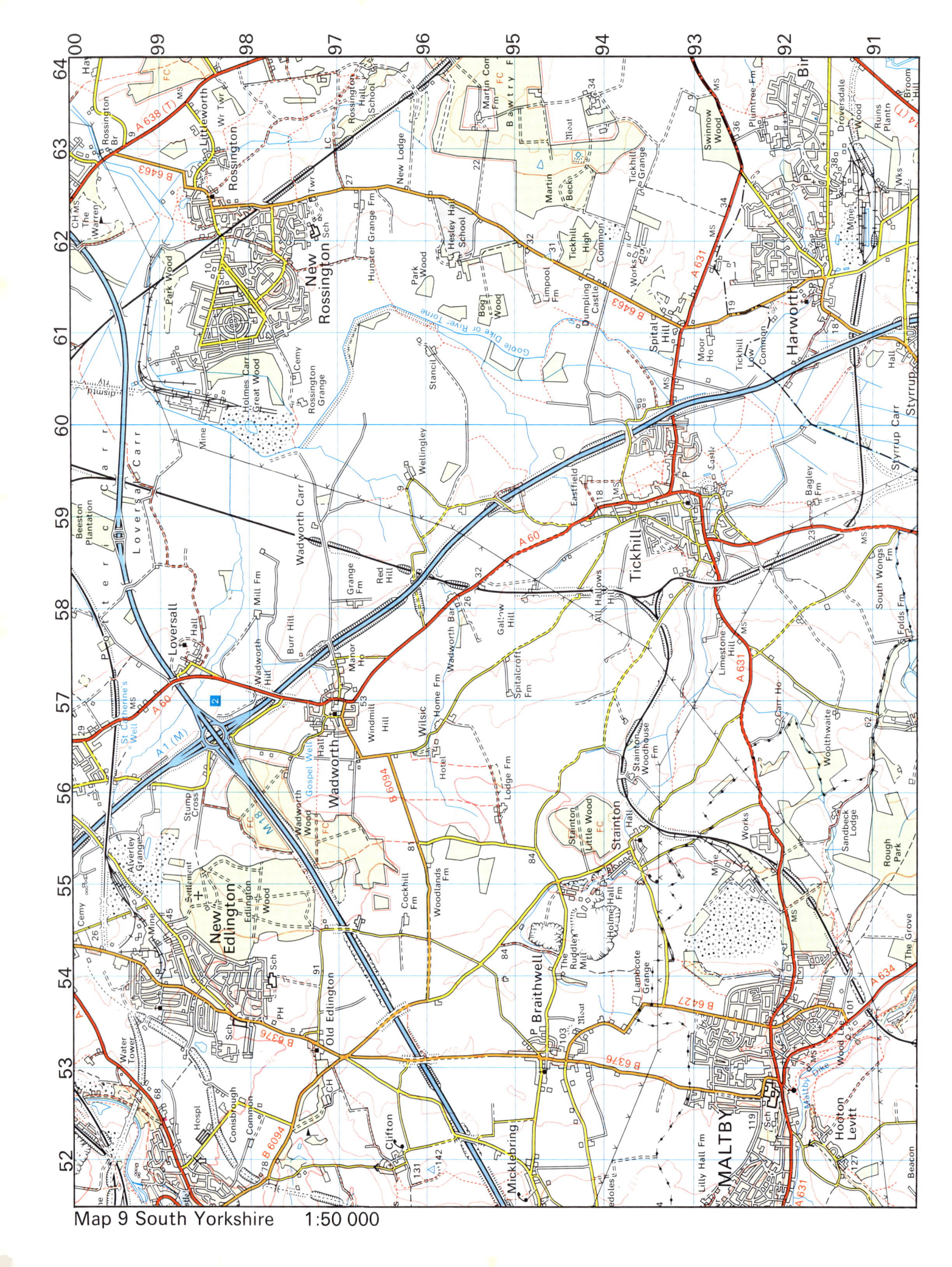

of this town lies below 20 m, the river channel having been straightened (5174, 5275/5375) and possibly embanked to prevent flooding. Similarly there is little residential development on the floor of the Darent Valley below the 20 m contour. Within Dartford, however, the river channel has been modified and its flow regulated (Lock 539751) to prevent flooding. North of Dartford the landscape has been artificially drained (e.g. 5477; 5377; 5376) and the Darent is heavily embanked to its confluence with the Thames (540780).

There is little surface drainage indicated elsewhere on the map, despite the existence of well-developed valleys. The courses of the two major streams, a left-bank tributary of the Cray (confluence at 5074) and that flowing north from Sundridge Park (4170), have been interrupted over short distances by residential development and the construction of roads and railways.

8.4 Describe the land-use patterns within 1 kilometre of either bank of the River Cray between St. Mary Cray (470675) and its confluence with the River Darent (536760). Illustrate your answer with a sketch map showing the distribution of the following land uses: residential; industrial; recreational; agricultural/ horticultural; transport and communications.

8.5 Repeat exercise 8.4 for the Darent valley between the road bridge at Farningham (546670) and the Thames confluence at 540780. What major differences do you observe between the land uses in the Cray and Darent Valleys?

Plate 8

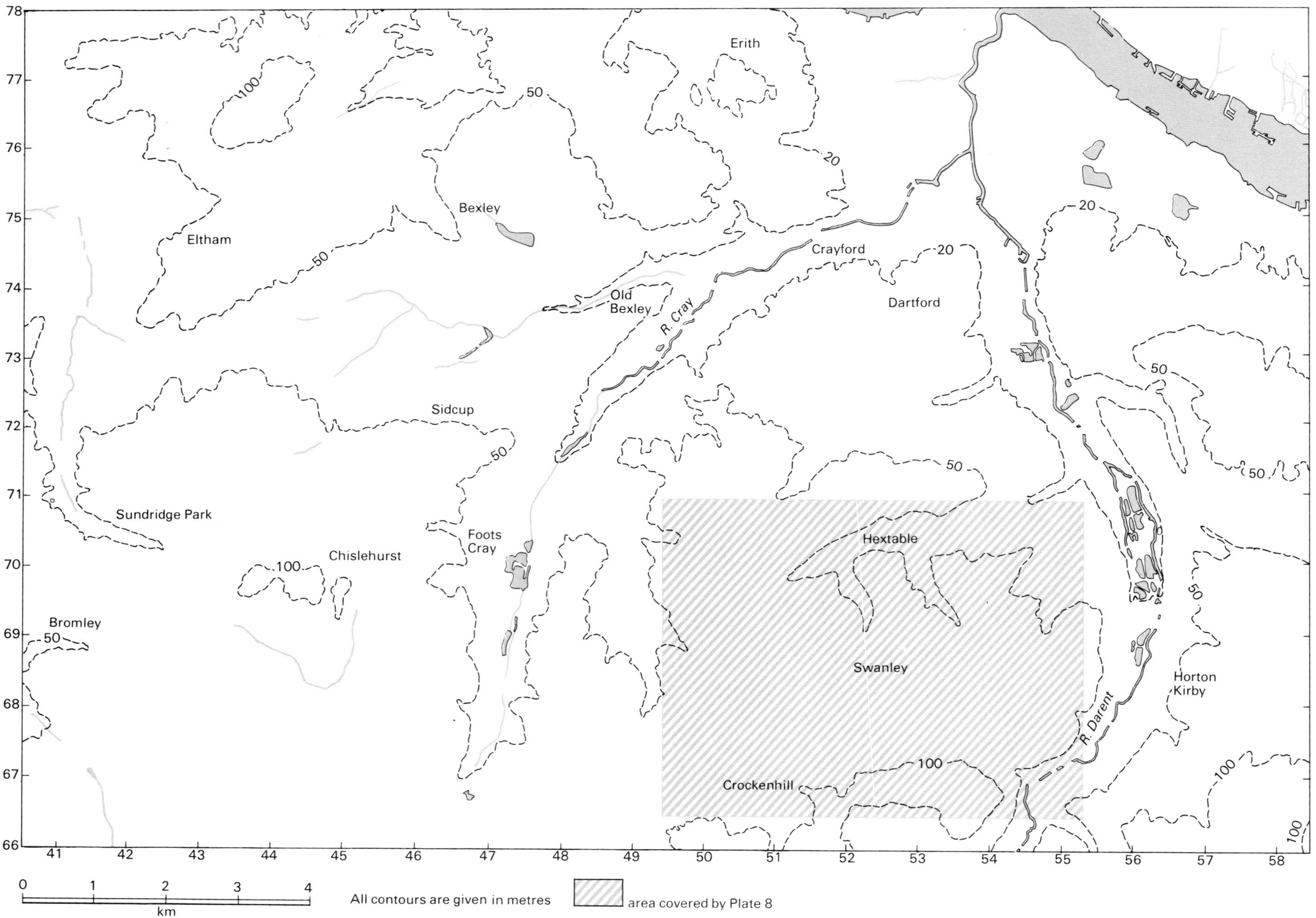

Fig. 8.2 Relief and drainage

8B.3 · Land use transect

Figure 8.3 summarises the contrasting densities in urban land use noted above. The cross sections show clearly the reduced amount of urban land use in the Cray and Darent valleys. They also highlight the importance of the parks and woodland in punctuating an otherwise continuous sprawl of settlement west of the River Cray. Figure 8.4 complements Fig. 8.3A and identifies the range of land uses found in each kilometric grid square between Northings 72 and 73.

8.6 Refer to Fig. 8.3. Draw similar, annotated, sections to show the density of the built-up area between the following pairs of Eastings: 43–44; 52–53; 57–58. What conclusions can you draw from these diagrams?

8.7 Study Figs. 8.3A and 8.4.
(i) Explain why there are so many major road junctions to the east of Easting 50.
(ii) Comment on the spatial distribution of (a) churches, (b) schools, (c) hospitals.

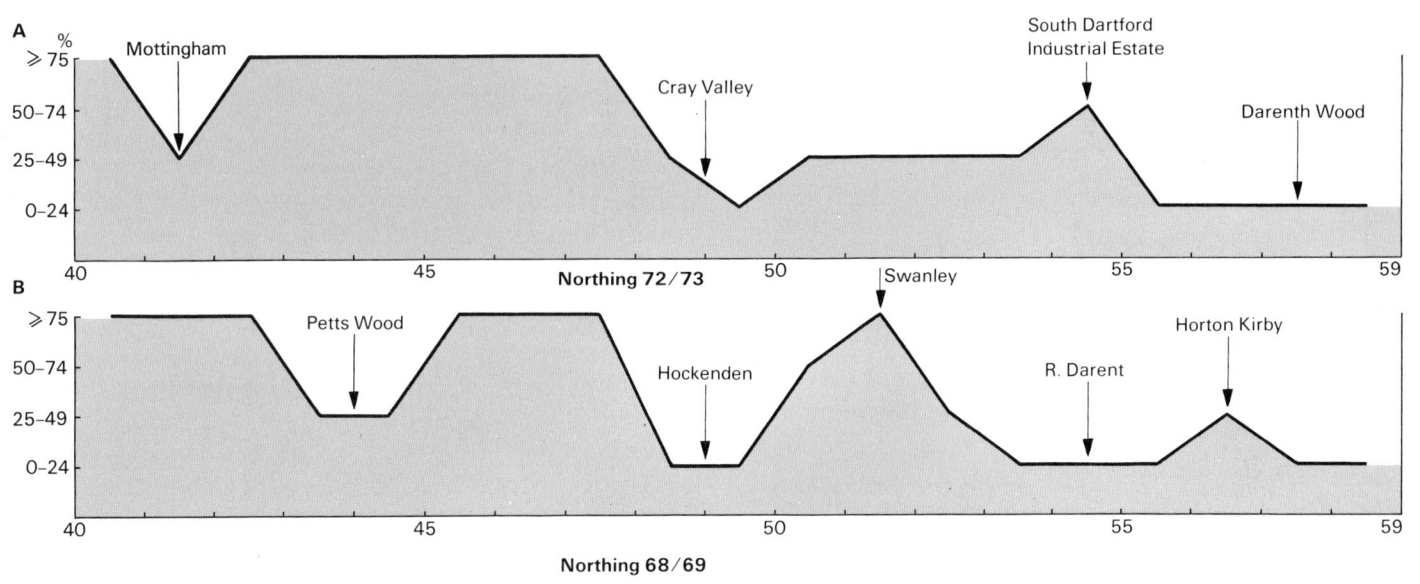

Fig. 8.3 Percentage of each kilometre grid square covered by residential development

Land use	40	41	42	43	44	45	46	47	48	49	50	51	52	53	54	55	56	57	58	59
A Residential (H)	H	H	H	H	H	H	H	H	H	H	H	H	H	H	H	H	H	H	H	H
B Communications																				
Railway (R)	R			R	R	R	R	R												
Station (St)	St						St	St												
Dual carriageway (D)			D	D	D					D	D		D	D	D	D	D	D	D	D
Major road junction (J)												J				J				J
C Institutions																				
Church (C)	3C	2C	4C		2C		2C	C					2C		C			C		
School (S)			2S				2S	S												
College (Co)		Co												Co						
Hospital (Ho)	Ho										Ho						2Ho	Ho		
D General urban																				
Industry (I)															I					
Crematorium/ cemetery (Ce)	Ce						Ce													
Golf course (G)															G					
Park (P)									P											
E Rural																				
Agriculture (A)											A	A	A	A	A	A	A	A	A	A
Glasshouse (Gl)													Gl							
Woodland (W)							W	W	W	W		W		W		W	W	W		W

Fig. 8.4 Principal land uses between Northings 72 and 73

How do these relate to the percentage of the land surface covered by residential uses?

(iii) Describe, with the aid of sketch maps, the assortment of land uses found in the following grid squares: 4072; 4872; 5472; 5772.

8.8 Compare and contrast the hospitals found in grid squares 4671, 5476 and 5772 with respect to (i) size (ii) location (iii) access to major roads (iv) general environment.

8.9 Adopt the land-use categories used in Fig. 8.4, and draw annotated transects between Northings 75–76 and 66–67. Compare these transects with Fig. 8.4. What conclusions do you draw regarding land-use patterns on the urban fringe?

8C · Land use on the urban–rural fringe

Figure 8.5 illustrates the changes in land use which may result from the expansion of a city into the surrounding countryside. These ideas are applied to the metropolitan fringe in South East London. Much of this area was built up in the interwar period and its expansion has continued since 1945. Map evidence is selected to demonstrate the following changes which have occurred in the rural landscape.

The absorption of rural settlements to form part of the built-up area. Eltham provides a good example of this process (Fig. 8.6). In historical times the village was centred around Eltham Palace (425740). From the arrangement of roads it seems that the main street is aligned east–west in grid squares 4274 and 4374. Away from this axis the rectangular street patterns in 4375 and newer housing estates with curvilinear networks of roads indicate different ages of housing development (See Chapter 7, 7C.1). Figure 8.6 shows that a range of services and facilities is available for residents. There are three railway stations and a good network of A class roads. Equally important, there are several areas of land at the edge of the built-up area which have been preserved for the public as open space.

8.10 On tracing paper mark the outline of the built-up area of Dartford, including the housing estates of Temple Hill (5575) and Fleet Downs (5673). Use initial letters to mark the sites of all hospitals, schools, golf courses, public open spaces, cemeteries and industrial land uses within the built-up area and within one kilometre of its

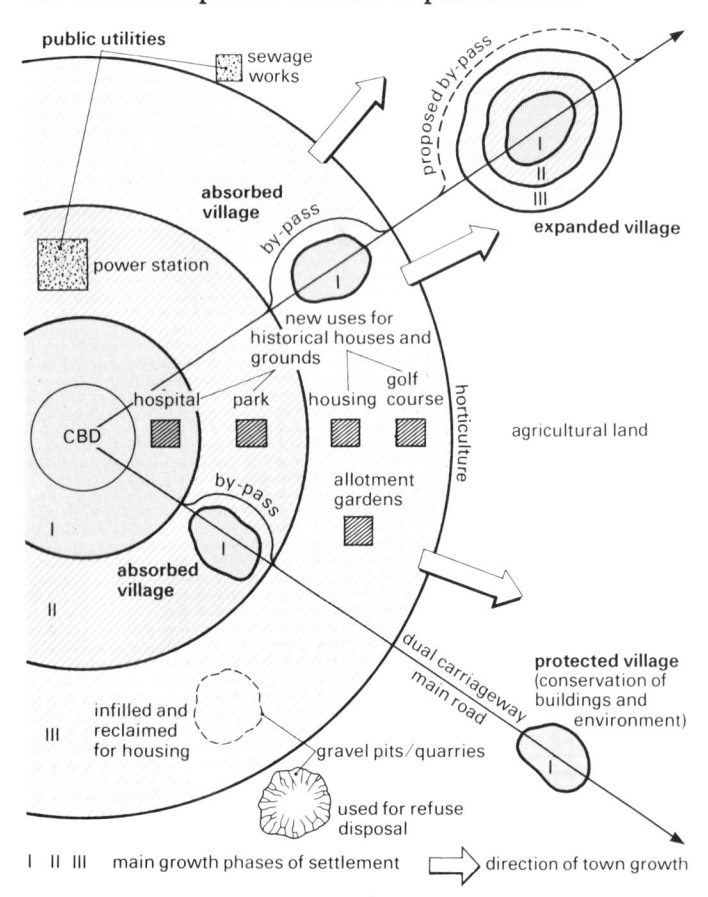

Fig. 8.5 Urban growth and changes in rural land use

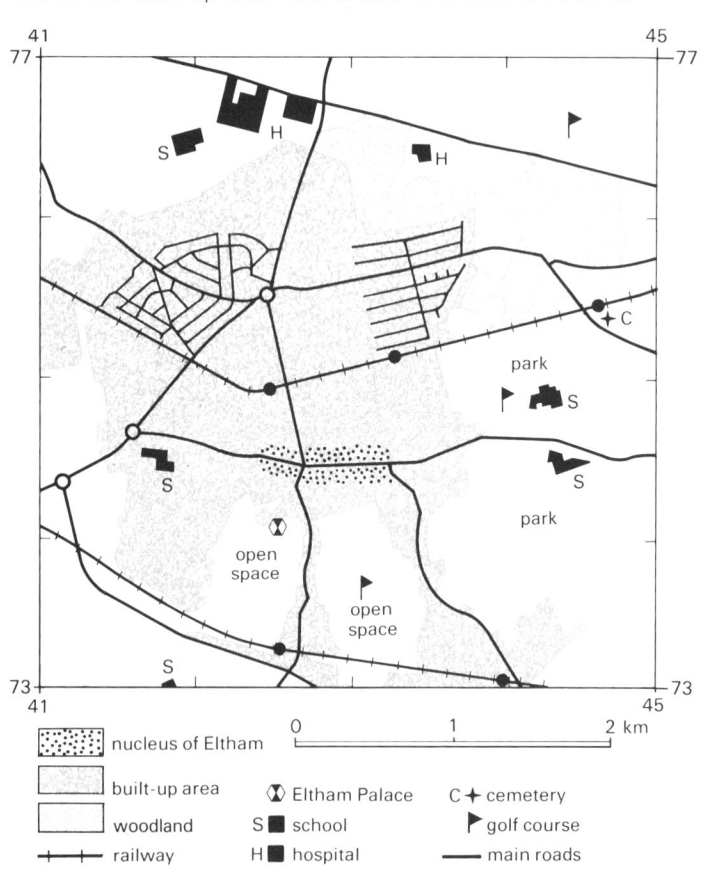

Fig. 8.6 The growth of Eltham

edge. Compare this diagram with Fig. 8.6. What similarities do you note between the land-use patterns of Dartford and Eltham?

8.11 Refer to Map 7 (Lower Teesside). What evidence can you assemble from the map to suggest that Ormesby (5317) was once a separate settlement?

The growth of 'satellite' villages. The development of the road and railway network provides an opportunity for some of those employed in the city to live in neighbouring settlements. Those villages most accessible to railway stations and having access to good roads will therefore be attractive to urban commuters. Ever since the 1930s, when planners designated large areas of the countryside around London as a Green Belt, there has been a land-use policy to check urban sprawl and to prevent neighbouring towns from merging into one another. Planning authorities therefore exercise a strict control over the location and amount of new house-building in the Green Belt. While housing developments are permitted in some villages, other villages may be protected as conservation areas on account of their natural beauty or architecture.

Swanley is a settlement which has experienced rapid growth. Swanley Village (5269), the historical core, shows little evidence of housing development. In contrast, there has been widespread residential development in the vicinity of the railway station (509682) and the crossroads (516686). This planned expansion of the village has consequently justified the construction of a by-pass which skirts the southern edge of the settlement.

8.12 Draw a simplified land-use map of Crockenhill (5066). What, in your opinion, are the attractive features of this village as a place in which to live? For what reasons might a property developer wish to build an estate of 500 houses for commuters to London?

8.13 Study Map 8 and Plate 8 and suggest the main occupations of the people living in Crockenhill.

Improvements in the road network. Increased volumes of road traffic converging on important cities have naturally increased the amount of congestion in the settlements which lie astride the main roads. This problem has been tackled in two main ways: firstly, the construction of by-passes around serious bottlenecks; secondly, the upgrading of A class roads into dual carriageways.

There is evidence on the A20 of both these solutions to traffic problems. By-passes have been constructed around Farningham (551667 – 543673), Swanley (525678 – 495698) and Sidcup (479707 – 449723) to relieve traffic congestion in the main streets of these historical settlements. Sections of this road, too, have been up-graded to the status of a dual carriageway (408746 – 450723; 455713 – 462708; 465705 – 471704; 495699 – 535674).

8.14 Explain why Farningham needed a by-pass. Refer to the map and air photograph and describe the route taken by this by-pass. Use a sketch map to illustrate local gradients and the method of road construction.

8.15 Study the settlements of Ormesby (5317) and Wolviston (4525) on Map 7 (Lower Teesside). Examine the changes in the road network that have taken place, or are taking place, within 1.5 km of the church in each settlement. Give reasons for these changes.

The loss of agricultural land. Extensive areas of agricultural land have been taken for residential, industrial and recreational purposes in South East London. Recreational areas remain as parks and public open spaces in the inner suburbs (Figs. 8.1, 8.6). Although these losses of agricultural land are now carefully controlled by the planning authorities, isolated farms surrounded by housing (Tong Farm 4468; Manor Farm 4871) and the location of new housing estates in agricultural areas (e.g. Hawley 5471; Wilmington 5371; Leyton Cross 5272) show the continuation of this process.

As the town expands, its population will require supplies of gas, water and electricity. These public utilities will need land. So, too, will sewage treatment plants and refuse disposal sites.

8.16 Prepare a sketch map to illustrate the encroachment of urban land uses on agricultural land in grid squares 4368, 4468, 4369, 4469 on Map 8.

8.17 Explain what seems to have happened to the land of Prissick Farm (510165) on Map 7 (Lower Teesside).

8.18 Examine Map 9 (South Yorkshire), grid squares 6292 and 6392. What is happening to the agricultural land in the vicinity of Plumtree Farm on the northern edge of Harworth?

8.19 Identify the public utilities at the following points on Map 9 (South Yorkshire): 550999; 536907; 613922. What do these locations have in common?

Specialised agricultural crops. Urban populations provide a ready market for fresh fruit and vegetables. It is therefore common to find specialised areas devoted to the production of these crops near large towns. As Fig. 8.7 and Plate 8 show, there are considerable acreages of orchards and several clusters of glasshouses in the vicinity of Dartford, Hextable and Swanley. These regions are well-connected by road to the markets of suburban London.

8.20 Use Map 8 and Plate 8 to answer these questions on agricultural land use.
 (i) What land uses are found at points, F, G, H and J on the air photograph?
 (ii) Identify and name two farms, giving a map reference for each.
 (iii) Explain the contrasting tones of land use shown on the air photograph in grid squares 5468 and 5367.

8.21 Analyse Fig. 8.7. Relate the distribution of glasshouses and orchards to: (i) the physical geography of that area; (ii) road communications.

Allocation of land for recreational purposes.
Although sections of the urban population spend their leisure time in cinemas, sports halls and leisure centres, those who prefer outdoor pursuits may seek sports grounds, golf courses, parks and areas of woodland for their recreation. Planning authorities in urban areas are therefore conscious of the need for recreational facilities and public open spaces. As Fig. 8.8 shows, pockets of land within the suburbs are reserved as golf courses, parks or open common land (see also Fig. 8.1 and Fig. 8.6). In the surrounding countryside, too, Green Belt planning controls have ensured that areas of land are available for recreational purposes. Here, as Fig. 8.8 indicates, the National Trust and Forestry Commission admit the public to areas of woodland under their control (e.g. Petts Wood 4468; Joyden's Wood 4971). There are, too, several buildings of historic interest in the area which are open to the public (e.g. St. Johns 558704; Hall Place 503742; Eltham Palace 424740).

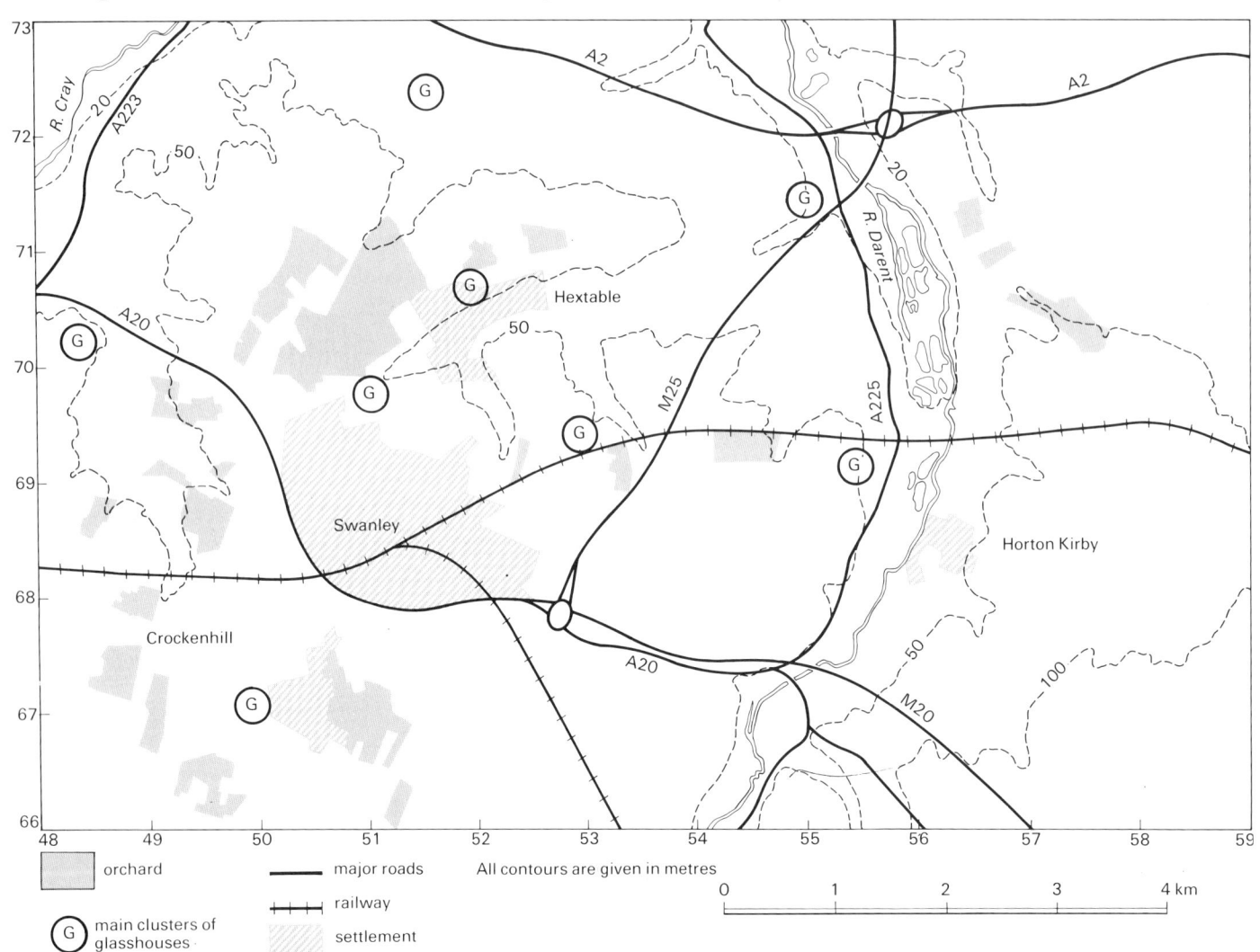

Fig. 8.7 Distribution of orchards and glasshouses

8.22 Examine Fig. 8.8. Count the number of golf courses and comment on their location. Compare this distribution pattern with that shown on Map 1 (Brighton Region).

Describe the characteristics of the residential areas closest to the golf courses in Brighton and Middlesbrough (Map 7). (Refer to Table 6.1.)

8.23 What recreational activities might take place on the lakes in the Darent valley?

8.24 Examine the land-use patterns on the margins of the built-up areas in Brighton (Map 1; 1:50 000) and Middlesbrough/Billingham (Map 7, 1:50 000). Compare the distribution, number, and kinds of recreational sites shown on these maps with those available on the fringe of South East London. What conclusions can you draw from this exercise?

There is widespread evidence to show that the use of land can change quite quickly on the fringe of urban areas. Different kinds of activity will often compete for the same sites; for example, housing with agriculture or industry with agriculture. This competition cannot be measured from the map, although the result of such changes can be seen. In some circumstances large mansion houses with spacious grounds are sold. The buildings may then become hospitals or institutions. These institutions often use their grounds for extending the buildings, or sell land for other purposes. In some cases, however, the land is reserved for recreational use.

8D · Road and rail communications

8D.1 · The basic network

Map evidence suggests that the growth of London has had a considerable impact on the

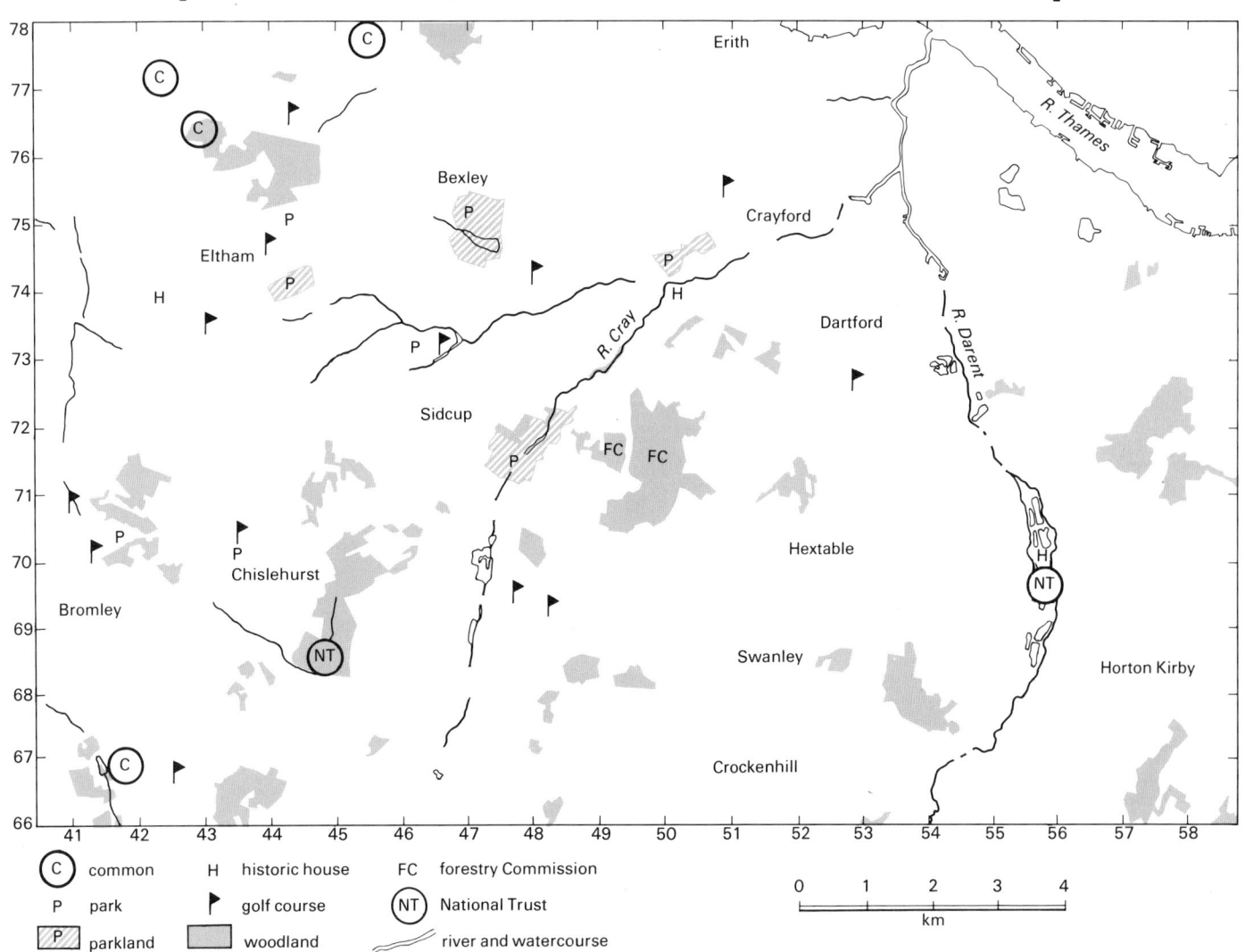

Fig. 8.8 Recreation and amenity land uses

pattern of road and rail communication in the region.

(i) The road system has been developed extensively since Roman times. It now comprises local service roads and major through roads. Several important roads have been upgraded into dual carriageways, by-passes have been constructed to reduce congestion, and motorways built as part of the national system.

(ii) A network of railways converging on central London has been constructed, mainly in the 19th century. This system links London with other towns in the south east; it also serves important suburbs like Eltham, Bexley, Sidcup and Bromley.

8.25 Prepare a topological diagram to show the network of A class roads and motorways in the region (see Chapter 4, Section 4C.4). Indicate the sections of dual carriageway and motorway. What does this diagram show regarding the main directions of the principal roads and the density of the network?

8.26 Draw a topological diagram to show the network of railway lines in South East London. Comment on the main direction of these lines, and the distribution and spacing of stations.

Repeat exercise 8.26 for the railway networks shown on Map 1 (Brighton Region) and Map 7 (Teesside).

8D.2 · Road communications

By-pass construction. Where several busy roads converge on a town centre, 'local' and 'through' traffic are mixed together. This results in traffic congestion and presents a serious problem for local residents. Figure 8.9A illustrates this situation. The construction of a by-pass (or ring road if it encircles the town) provides a solution. Sometimes, however, rapidly growing residential and industrial areas will leapfrog the by-pass (compare Figs. 8.9A and 8.9B). When traffic from these newly developed areas has direct access to the by-pass, further congestion may occur and the movement of through traffic is hindered. Where traffic flows are particularly heavy, this situation can lead to the construction of a second by-pass.

Route description. There are three major roads leading into London from the east and south east: (i) the A207/A226; (ii) the A2(T); (iii) the A20(T).

(i) The old Roman Road (4476) is the most northerly of these roads. It is straight for much of its length despite local changes in the height of the land surface. Designated the A226 at Dartford (5674), it becomes the A207 in Crayford (5174), and continues west across Shooters Hill

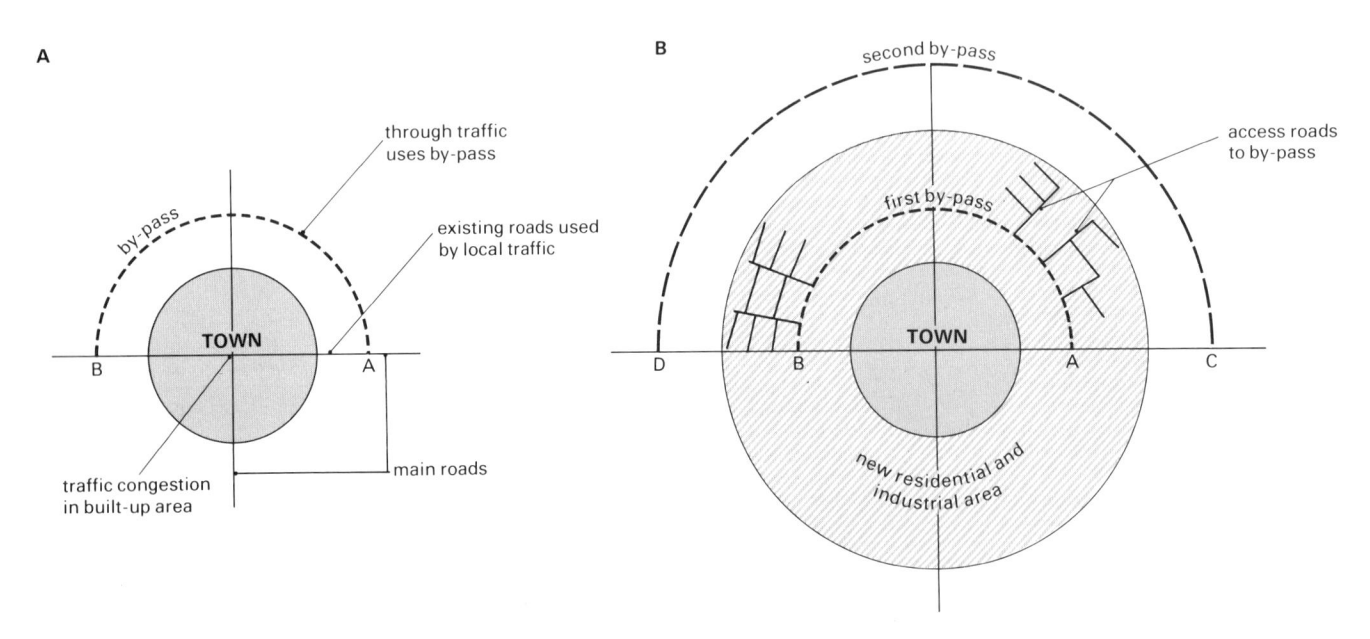

Fig. 8.9 By-pass construction and town growth

(4376) to London. Nowhere has this road been improved to the standard of a dual carriageway. The severe traffic congestion which resulted in the streets of Dartford and at the river crossing (5474) no doubt led to the construction of the first by-pass (Fig. 8.10). This road, the A296/A225 leaves the A226 at 569734 and rejoins it via the B2174 at 523746. Note, however, that a number of access roads from housing estates join this by-pass, for example in the vicinity of Fleet Downs (5673).

(ii) As Dartford grew this by-pass became congested. Traffic problems no doubt occurred, too, at the river crossing in Crayford (5174). This problem was solved in the interwar years by creating the A2. Now it is a major dual carriageway to serve the 'through' traffic destined for central London. Thus Dartford has both an 'inner' and 'outer' by-pass, as shown on Fig. 8.10. Judging by its appearance, the A2(T) has become the major radial road leading into south east London.

8.27 Refer to the guidelines on route description in Chapter 5 (page 61). Describe the route taken by the A207/A226 between 590748 and 400772. Compare this route with that following by the A2(T) between 590727 and 404780, with regard to: (i) the type and class of road, (ii) gradients encountered, (iii) likely points of traffic congestion, (iv) links with local road networks. Calculate the Detour Index for each road between these specified points. What conclusions do you draw from this exercise?

8.28 A motorist who lives alongside the school at Eltham Park (4474) wishes to visit a friend in hospital at Dartford (5773). Would you recommend the A2(T) or A226/A207 for this journey? Explain your decision assuming he wishes to start the journey at (i) 1400 hours on a Sunday, (ii) 1700 hours on a Friday.

8.29 Examine Fig. 8.10 carefully. Does the model of by-pass development shown in Fig. 8.9 help to explain the situation at Dartford? Give reasons to support your answer.

(iii) Further south, the A20(T) contains three major by-passes. The original A20(T) passed through the linear settlement of Farningham (5467). A recent by-pass lies to the north and uses an embankment and bridge to cross the Darent floodplain. Further west on the A20(T) there are important differences between the by-passes at Swanley (524679 to 498698) and Sidcup/Eltham (479706 to 412744). The Swanley by-pass, 4.0 km long, has access restricted to either end. In contrast, the Sidcup/Eltham by-pass, skirting the A211 via Ruxley, Sidcup, Longlands, New Eltham and Eltham, is 7.5 km long and has at least thirty access roads.

Restricted access is an important feature of motorway design. This can be seen on the M20 and M25, which are extensions of the A20(T) to the east and north.

8.30 Describe the route taken by the A20(T) between 559660 and 400749.
(i) Note major changes in gradient.
(ii) Seek evidence of by-passes.
(iii) Estimate the proportion of the road which is dual carriageway.

8.31 Refer to Fig. 8.10 and Plate 8. Draw a sketch map to show the by-passes at Farningham (5466) and Swanley (5268).

8.32 Suggest solutions to the traffic problems which occur at the following locations: 479707; 412743.

8.33 Use your knowledge of London's traffic problems to comment on the direction and grade of the road link between 575780 and 527678. What is the significance of the major road intersection in grid square 5367?

8.34 Estimate the area of land occupied by motorway junctions on (i) Map 8 (S.E. London) (ii) Map 9 (South Yorkshire).

8.35 From your experience, answer the following questions.
(i) To what kinds of traffic are motorways restricted?

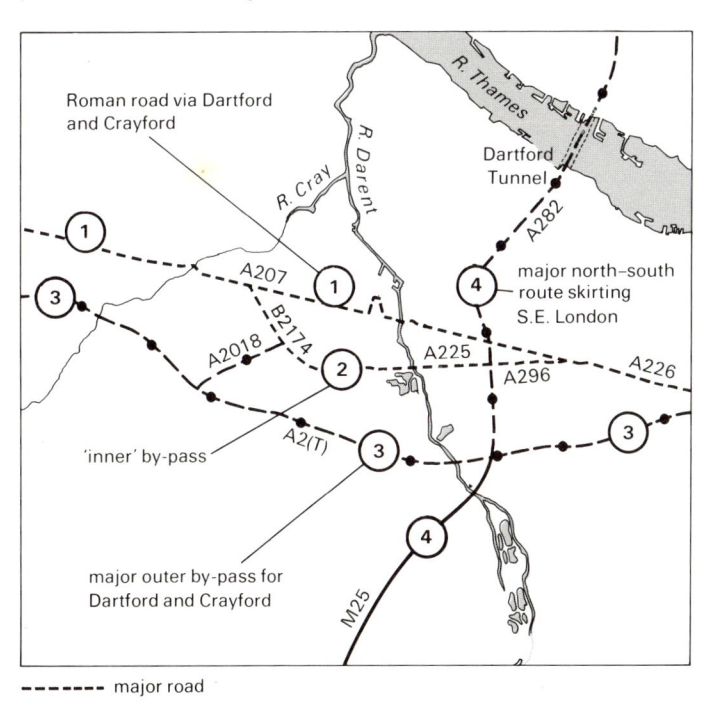

Fig. 8.10 Sketch map to show by-pass construction around Dartford/Crayford

(ii) How many carriageways are provided on motorways?
(iii) What access is there to motorways?
(iv) Do any buildings have direct access to motorways?
(v) How do construction engineers deal with the physical geography of areas through which they intend to build a motorway? Illustrate your answer with reference to Map 8 (S.E. London) and Map 9 (South Yorkshire).
(vi) What are the differences between motorways and A class trunk roads?

8.36 Plate 8 was taken before the construction of the M25. Describe the route taken by the M25 between the junctions in grid squares 5267 and 5572. What effect has the M25 had on the landscape through which it passes? Draw a sketch map to show how the M25 relates to the existing network of local roads.

8.37 Imagine that you are the managing director of an engineering firm located on the industrial estate in grid square 5472. You supply components, by road, to a number of customers in East London, South Wales and Teesside. Comment on the road network shown on Map 8 as it might affect your business.

8.38 All main routes except the M25 and its extension to the Dartford Tunnel trend east–west. Why is this? What transport problems would you expect to find if you wanted to travel from Erith (5077) to Sidcup (4672)?

8D.3 · Railway communications

Write a short statement on the railway system in South East London. As a basis for this account, answer the following questions.

8.39 Has the physical geography of the region presented any difficulties for the construction of railways? Give evidence to support your answer (Refer to Fig. 8.2).

8.40 Describe the routes taken by the railways between:
(i) Blackheath Park (4075) and Stone (5774);
(ii) the station at 401743 and Dartford (543745).
What do you observe regarding the spacing of the stations along these lines? Comment on the engineering problems involved when building these lines. Draw a sketch to show those sections of the lines found: (a) in cuttings; (b) on embankments.

8.41 What is the average distance separating the railway stations on the lines east of the River Cray? Why are these stations so close together?

8.42 List the main advantages and disadvantages facing those who commute daily by rail from stations in the suburbs to employment in central London.

8.43 What happens to the railway line at point K on the air photograph? Give a map reference for this.

CHAPTER 9
South Yorkshire

Chapter Plan

9A · **Physical landscape**
  9A.1 Relief and drainage
  9A.2 The transect: a basis for regional division

9B · **Economic activity and the rural landscape**
  9B.1 Industry and agriculture
  9B.2 Communication networks

9C · **The study of rural settlement**
  9C.1 Methods of study
  9C.2 Settlement in the study area

9A · Physical landscape

9A.1 · Relief and drainage

The procedures recommended in Chapter 1, Section 1D.1, were used to delimit three physical regions on Map 9. The contour trace (Fig. 9.1), line section (Fig. 9.2) and transect diagram (Fig. 9.3) illustrate the surface characteristics of these regions.

Region I, an upland area, covers almost half the map. The height of the land falls to the east, from 142 m (519959) to 15 m in approximately 7 km. Almost everywhere the topography is generally smooth and rounded. Several named 'hills' between 15 m and 46 m in height indicate the eastern margin of this upland area (Wadworth Hill 5797; Burr Hill 5797; Windmill Hill 5696; Gallow Hill 5795; All Hallows Hill 5894). In the western and higher part of this region there is no surface drainage. This may indicate that the underlying rock is limestone. Further east, however, streams flow north-eastwards in shallow valleys to the River Torne.

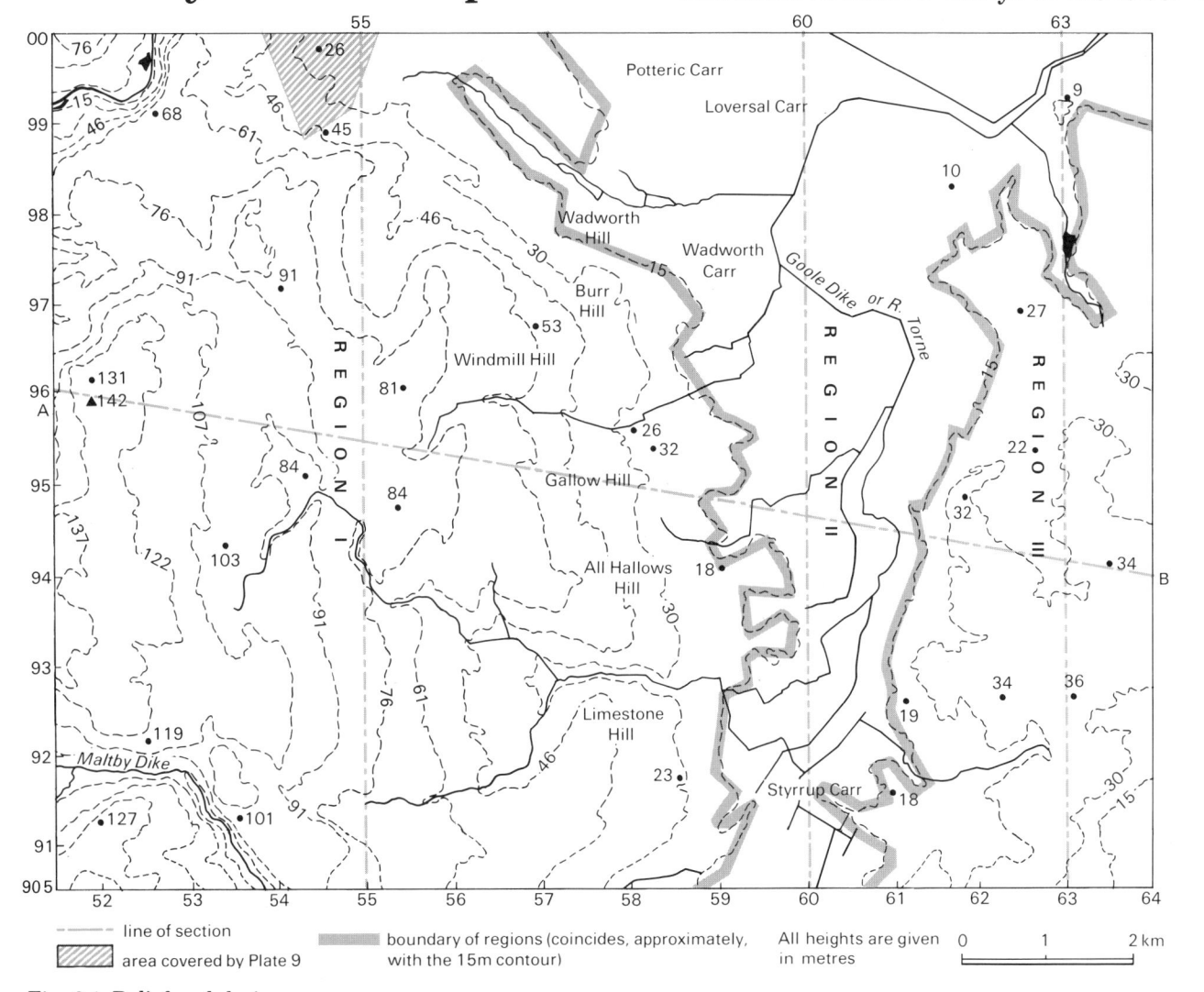

Fig. 9.1 Relief and drainage

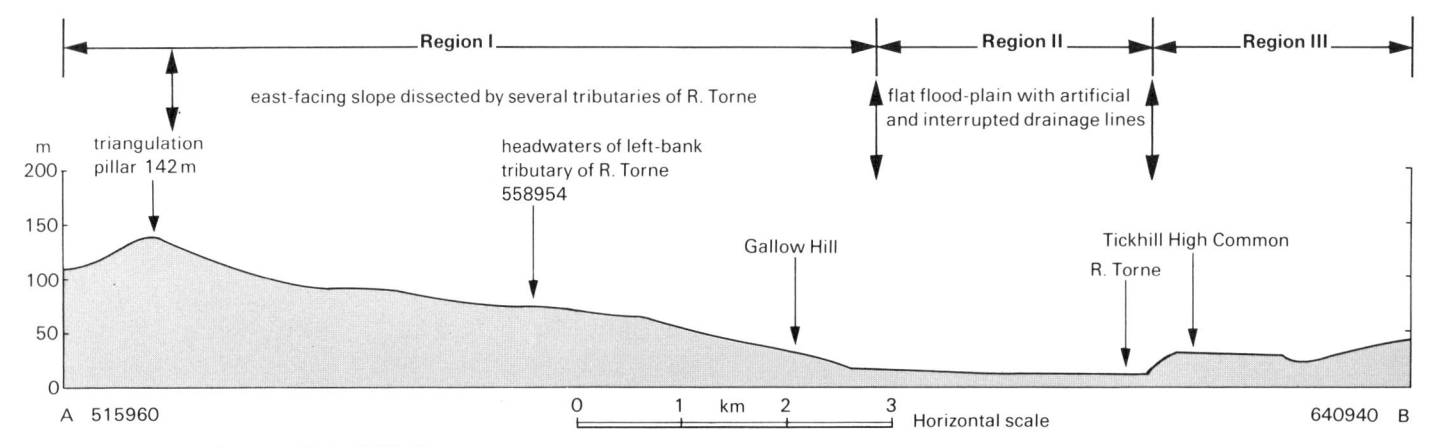

Fig. 9.2 Line section 515960–640940

Region II comprises the broad and relatively flat floor of the Torne Valley. For convenience the 15 m contour is taken as the general boundary of the region. The word 'carr', printed in several places, indicates a marshy and low-lying area, with gentle gradients (5899/5999; 5897/5997; 590905/600905). In some places the network of surface drainage has been artificially controlled. This results from mining activity, road construction, and farming practices.

(i) Sections of the River Torne, west of New Rossington, have been straightened between 610950 and 635000.

(ii) South west of New Rossington an embankment, 2.25 km long, has been built on the left bank of the River Torne (610959–597974).

(iii) Parts of several left-bank tributaries have been straightened, and other water courses diverted, in the process of road construction (5992; 6092; 6091; 5895).

(iv) Some water courses crossing agricultural land have also been artifically controlled (5998; 5996; 6094; 6095).

Region III, in the east, covers less than one fifth of the area. Its surface undulates gently between 15 m and 35 m. In this region there is little surface drainage, except to the west of Harworth (6191). The situations of Tickhill High Common (6194/6294) and Tickhill Low Common (6092) confirm the relative difference in relief between this region and the floodplain of the Torne.

9A.2 · The transect: a basis for regional division

Do the three physical regions have contrasting patterns of settlement, communications and economic activity? This question can be answered in two ways:

(i) By tracing each distribution from the map. These distributions can then be compared with the diagram showing the boundaries of the physical regions.

(ii) Alternatively, selected features in the physical and human landscapes can be summarised in a transect diagram. Figure 9.3 illustrates this procedure. It uses observations made in Region I, within 0.5 km of either side of Easting 55. Examine this transect carefully and note the headings used for collecting information from the map.

Prepare similar transects centred on Easting 60 for Region II, and Easting 63 for Region III.

9.1 Compare the distributions shown in these three transects. Write a short account of the main differences in the landscape between Regions I and II, and Regions II and III.

9B · Economic activity and the rural landscape

9B.1 · Industry and agriculture

Chapter 7 examines the development of industry and urban settlement in Teesside. In rural Britain, too, industrial developments have had a marked effect on the landscape in some areas.

Figure 9.4 indicates areas of industrial land use in the Tickhill region. These were delimited using the guidelines in Chapter 7 (7B.1). This exercise was quite easy given the indicators 'mine', symbols for waste tips and quarries, and networks of mineral lines. It is not possible,

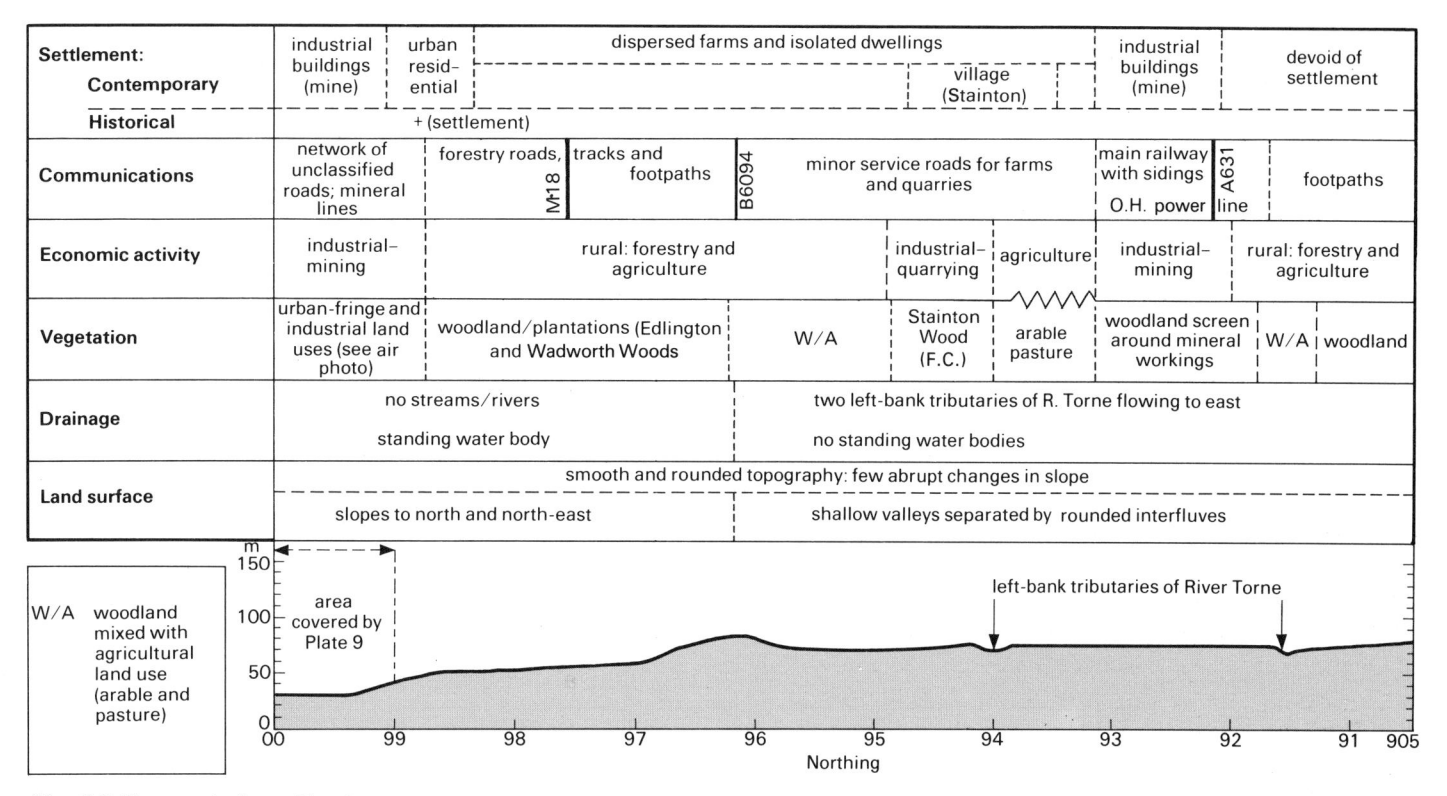

Settlement: Contemporary	industrial buildings (mine)	urban resid- ential	dispersed farms and isolated dwellings			industrial buildings (mine)	devoid of settlement
Historical		+ (settlement)			village (Stainton)		
Communications	network of unclassified roads; mineral lines	forestry roads, M18	tracks and footpaths B6094	minor service roads for farms and quarries		main railway with sidings A631 O.H. power line	footpaths
Economic activity	industrial- mining	rural: forestry and agriculture		industrial- quarrying	agriculture	industrial- mining	rural: forestry and agriculture
Vegetation	urban-fringe and industrial land uses (see air photo)	woodland/plantations (Edlington and Wadworth Woods)	W/A	Stainton Wood (F.C.)	arable pasture	woodland screen around mineral workings	W/A woodland
Drainage	no streams/rivers standing water body			two left-bank tributaries of R. Torne flowing to east no standing water bodies			
Land surface	smooth and rounded topography: few abrupt changes in slope						
	slopes to north and north-east			shallow valleys separated by rounded interfluves			

W/A woodland mixed with agricultural land use (arable and pasture)

Fig. 9.3 Transect along Easting 55

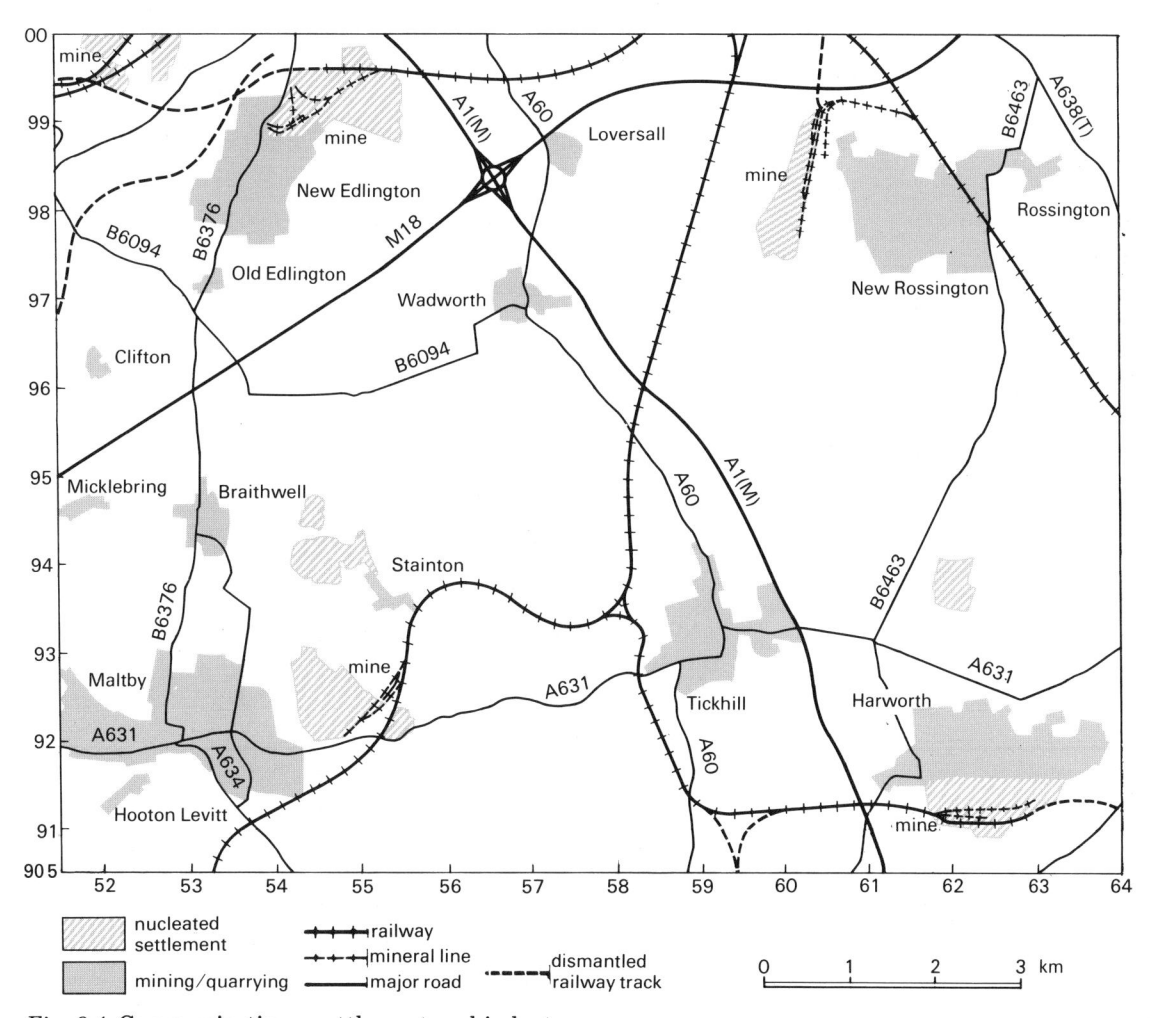

Fig. 9.4 Communications, settlement and industry

however, to identify the kinds of industry occupying the 'works' at 627905, 617936 and 555922.

Agriculture is the major land use outside the nucleated settlements. There is a fairly even distribution of named farms in Region I and Region II. These are probably engaged in dairying given the altitude, heavy soils and nearness to urban markets. Some arable crops are probably grown away from the flood-prone areas in Region II. There are extensive areas of woodland, some managed by the Forestry Commission. This land use represents a distinctive feature in the landscapes of Region I and Region III. In several localities, too, spoil tips and mining installations have been screened by belts of trees (e.g. 5492; 6097 – 6098).

9.2 What proportion of the map area is covered by mining waste?

9.3 What is the area of Edlington Wood (5497)?

9B.2 · Communication networks

Section 5C.1 in Chapter 5 provides a useful background for studying the networks of roads and railways in the vicinity of Tickhill. Read this carefully, and answer the following questions.

9.4 Study the route of the A1(M) between 612905 and 555000, and answer these questions.
 (i) How many bridges cross the motorway?
 (ii) Specify the kinds of road and/or railway that these bridges support.
 (iii) How many roads and railways are crossed by the motorway?
 (iv) What proportion of the total length of the motorway is (a) in cuttings, (b) on embankments?
 (v) Why does a motorway need so many cuttings and embankments?
 (vi) What area of land is taken up by the motorway junction in grid square 5698?

Answer questions (i) to (iv) for the M18 between 515951 and 620000.

9.5 Identify the feature represented on the M18 at 588995. What does this imply for the future pattern of roads in the region?

9.6 Has the M18 seriously affected the beauty of the landscape through which it passes? Give evidence from the map to support your viewpoint.

9.7 Prepare a tracing overlay of the map and mark railway lines, dismantled railway lines and mineral lines.

Now answer the following questions.
 (i) What proportion of the total length of the original main line railway track (exclude mineral lines) has been dismantled?
 (ii) Suggest reasons why sections of the railway track have been dismantled.
 (iii) What purpose is served by the mineral lines?

9.8 With regard to the physical geography, describe the route followed by the railway between 533905 and 596000. What proportion of this route is (i) contained in cuttings, (ii) supported on embankments? Why are there so many cuttings and embankments along this route?

9.9 Explain why the railway describes an 'S' bend between Easting 55 and Easting 59.

9.10 Describe the route followed by the A631 between 515920 and 640931.

9.11 Refer to Fig. 9.1. Describe the pattern of road and railway communications in Region II.

9C · The study of rural settlement

9C.1 · Methods of study

There are three important characteristics of rural settlements which can be studied from an Ordnance Survey map. These are:

(i) the spatial pattern of settlements in a prescribed area;

(ii) studies of the site, situation, function and shape (morphology) of individual units of settlement;

(iii) classification of settlement types.

All three approaches involve a study of detail on an Ordnance Survey map. The amount of information given varies with the scale of the map. Table 9.1 shows, however, that the Ordnance Survey uses different sizes of print-face to give general guidance on the size and kind of settlement units.

9.12 Examine the following grid squares:
Map 6   Central Bath   1:10 000   7564
Map 4   Wharfedale   1:25 000   9672/9772
Map 1   Brighton Region   1:50 000   3104
Compare and contrast the detailed representation of settlement in these areas.

Spatial pattern. There are three ways in which the pattern of settlement can be simplified, and then analysed:

Table 9.1 Types of settlement shown on Second Series 1:50 000 metric maps

Settlement type	Example and location		
Towns			
Over 20 km² in area	**MIDDLESBROUGH**	(4918)	Map 7
5 – 20 km² in area	**BEXLEY**	(4675)	Map 8
Under 5 km² in area	BETHESDA	(6266)	Map 3
Suburban areas	New Eltham	(4472)	Map 8
Villages			
Over ½ km² in area	Hextable	(5270)	Map 8
Under ½ km² in area	Swanley Village	(5369)	Map 8
Hamlet	Grubb Street	(5969)	Map 8
Farm	Mussenden Fm	(5667)	Map 8

(i) a transect diagram, as for example in Figure 9.3;

(ii) an annotated sketch map which relates characteristic types of settlement to physical regions in the landscape, as for example on Fig. 1.16;

(iii) the use of words like 'dispersed' or 'nucleated' to describe the spatial distribution of settlement units in the landscape (see Fig. 9.5). There are, too, simple statistical methods which can be used to measure the degree of dispersion or clustering in the settlement pattern (see B.D.R. Worthington and R. Gant: *Techniques in Map Analysis*, Macmillan 1975, pages 12–15).

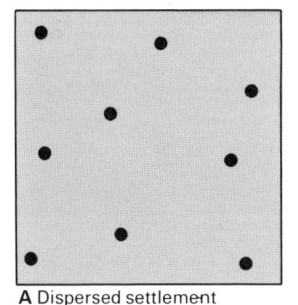

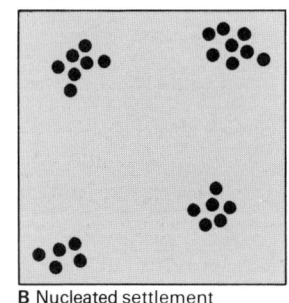

A Dispersed settlement   **B** Nucleated settlement
C Mixed nucleated and dispersed settlements

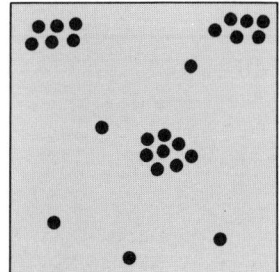

Fig. 9.5 Settlement distribution

Representative types of rural settlement found in Britain are shown on Fig. 9.6.

9.13 Refer to Map 2B (Stour Valley) and Fig. 9.5. Use the appropriate terms to describe the settlement pattern in the following areas:
 (i) between Northings 025 and 090 and Eastings 800 and 860;
 (ii) between Northings 090 and 150 and Eastings 870 and 890.

Study of individual settlement units. A more detailed examination of a settlement unit can be carried out using the guidelines set out below for site, situation, shape and function. These headings can also provide a basis for classifying different types of settlement.

(i) *Site*: the area of land covered by the buildings which make up the settlement, together with the related open spaces, roads and railway tracks inside the perimeter of the built-up area.

(a) relative relief i.e. height of settlement relative to the immediate surrounding area;

(b) slope: gradients within the site;

(c) drainage;

(d) aspect;

(e) margins of the settlement: relationship to physical features.

(ii) *Situation*: the site of the settlement with regard to the area covered by the map.

(a) Physical setting: relative relief (see above), drainage, slope, geology;

(b) communication network: road, rail, canal,

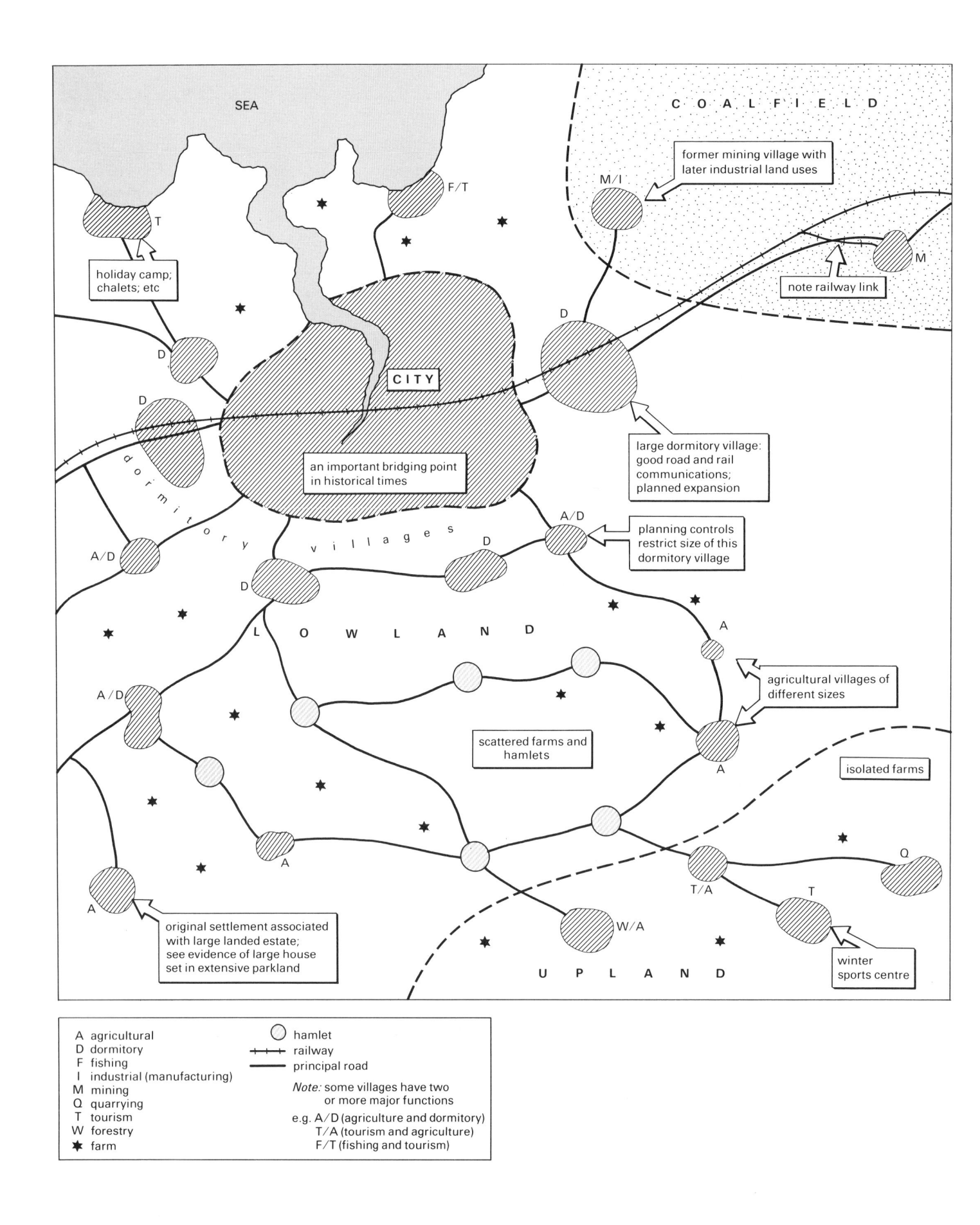

Fig. 9.6 Representative types of rural settlement

106

class of roads (e.g. A, B, etc) rail (single, double track, etc);

(c) administrative setting: parish; administrative districts.

(iii) *Shape (morphology)*: the shape and arrangement of the settlement e.g. street pattern; land-use characteristics; relationship to site.

(a) Buildings: pattern; block size and shape; named uses;

(b) internal road and railway network: classification of roads; density of network; nodal points;

(c) open space: spaces around buildings; kinds of open spaces;

(d) historical buildings.

(iv) *Function (economy)*: the activities and services performed by the settlement for its population, and other people living outside.

(a) Use of buildings: residential; civic; service; industrial; agricultural;

(b) spatial groupings of services and activities.

9.14 Consider these guidelines carefully, and write a short account of the site, situation, shape (morphology) and functions of: (i) Tickhill (Map 9, 5893), (ii) Poynings (Map 1, 2612).
How does Plate 1A help with the study of Poynings?

9.15 On the basis of their shape, most nucleated settlements can be classified as being either linear or clustered. Decide which of these general labels best describes the following settlements on Map 2B (Stour Valley).

Name	Grid reference
Winterborne Stickland	8304
Iwerne Minster	8614
Child Okeford	8312
Shillingstone	8211
Stourpaine	8609
Bryanston	8707
Durweston	8508

What problems did you find in completing this exercise?

Classification of settlement units

9.16 We often describe a settlement by features of its site, shape, history or economy. Examine the lists below, and find a suitable label from List 2 for each settlement in List 1.

List 1 Settlement and location

Name	Map reference	
New Rossington	Map 9	6197
Fulking	Map 1	2411
Coed y parc	Map 3	6166
Tickhill	Map 9	5893
Swanley Village	Map 8	5269
Starbotton	Map 4	9574
Wolviston	Map 7	4525
South Pool	Map 5	7740
Woodmancote	Map 1	2314
East Prawle	Map 5	7836
Nant Peris	Map 3	6058
Salcombe	Map 5	7338
Kingsbridge	Map 5	7344
Middlesbrough	Map 7	5021

List 2 Settlement features

fording point, valley head, defensive point, bridging point, spring line, industrial estate, quarry, mining, ria head, lowest bridging point, tourist, ria mouth, commuter, agricultural.

9.17 What problems did you find in doing exercise 9.16? What do they tell us about the practice of classifying rural settlements?

9.18 Study Map 3, and answer questions 3.25 to 3.29 on page 40.

9C.2 · Settlement in the study area

Spatial pattern and types of settlement. From the study of the transects prepared for Eastings 55, 60 and 63 you will be familiar with the variety of settlement types shown on Map 9. This settlement distribution includes industrial villages (e.g. Maltby 5291/5491 – 5293/5493, Harworth 6191/6391 – 6192/6392), non-industrial villages (e.g. Tickhill 5892/6092 – 5894/6094, Braithwell 5394), hamlets (e.g. Old Edlington 5397, Loversall 5798) and isolated farms (e.g. Bagley Farm 5991; Limpool Farm 6194). Refer to the three transects and Fig. 9.1, and answer question 9.19.

9.19 With regard to size, spatial pattern, and type of settlement units, compare and contrast the settlement distributions in Regions I, II and III.

Historical growth of settlement. Three phases can be identified in the settlement history of the region from evidence on the map.

(i) As the transects show, the evidence for historical settlement is widely scattered and confined to Region I and Region III.

9.20 List the features of historical settlement in each of the three physical regions shown on Fig. 9.1

(ii) More recently, and before the intrusion of mining activity, there were two main elements in the settlement pattern: dispersed farms and small nucleations. The farms are scattered across the map, although their density is lower in Region II, immediately west of the River Torne. Hamlets and villages, clustered around a church, were the focal points in the network of local roads. Some, like Wadworth (5696) and Braithwell (5394), have historical origins. These nucleations would have provided services for the local agricultural communities.

Plate 9

9.21 Use the guidelines for the study of settlement set out in Section 9C.1 above, and write a short account of the site, situation, form and function of: Braithwell (5394); Wadworth (5696); Tickhill (5893).

(iii) Mining activity seems to have been directly responsible for the rapid growth of at least four settlements: New Rossington (6197), New Edlington (5498), Harworth (6191) and Maltby (5292). In each instance, large areas of housing have been built close to the mining installation. Note, too, that the parent settlement Old Edlington is situated approximately 0.75 km from the newer mining development of New Edlington. Similarly, the historical core of Rossington lies on the eastern margin of the new mining centre, New Rossington. Furthermore it seems probable that Stainton (5593) has developed in relation to the nearby quarries.

9.22 Measure the total built-up areas of the following settlements:
(i) New Edlington, (ii) New Rossington/Rossington, (iii) Maltby, (iv) Harworth. (Include residential and industrial land uses.) In each settlement work out the proportion of the land devoted to industrial uses. (Include mines, 'works' and spoil tips.) Comment on the figures you calculate.

9.23 Examine Plate 9 carefully. Prepare on tracing paper an overlay which distinguishes the following land-use regions on the fringe of New Edlington: (i) residential, (ii) industrial, (iii) communications, (iv) recreational.

9.24 Compare the railway network shown on the map and air photograph. What changes have occurred since the air photograph was taken?

9.25 What land uses are shown on the photograph by the letters A, B, C and D? Give map references for these features.

9.26 Assemble evidence from Plate 9 to demonstrate that the mine is still producing coal. As a guide, refer to those features indicated by the letters D, E, F and G.

9.27 Describe the shape and plan of the waste disposal area at B on Plate 9. Why do you think it is arranged in this form?

9.28 Examine the residential area in the south of the photograph and answer these questions.
 (i) What is the installation at H?
 (ii) Describe the main differences in the architecture and plan of the housing units in areas J, K and L.
 (iii) Rank the housing areas J, K and L with respect to their age.
 Give reasons for the order you choose.

The mining village: a land-use model.
Descriptions like 'spring-line' or 'ria-head' are used to identify villages found in a particular site. In a similar way, settlements can be classified according to a common pattern of land use. Examine Fig. 9.7, a simplified sketch of land-use regions in the mining settlement of Harworth. Three major land-use regions are distinguished:

(i) a historical core, on the western margin of the present-day built-up area;

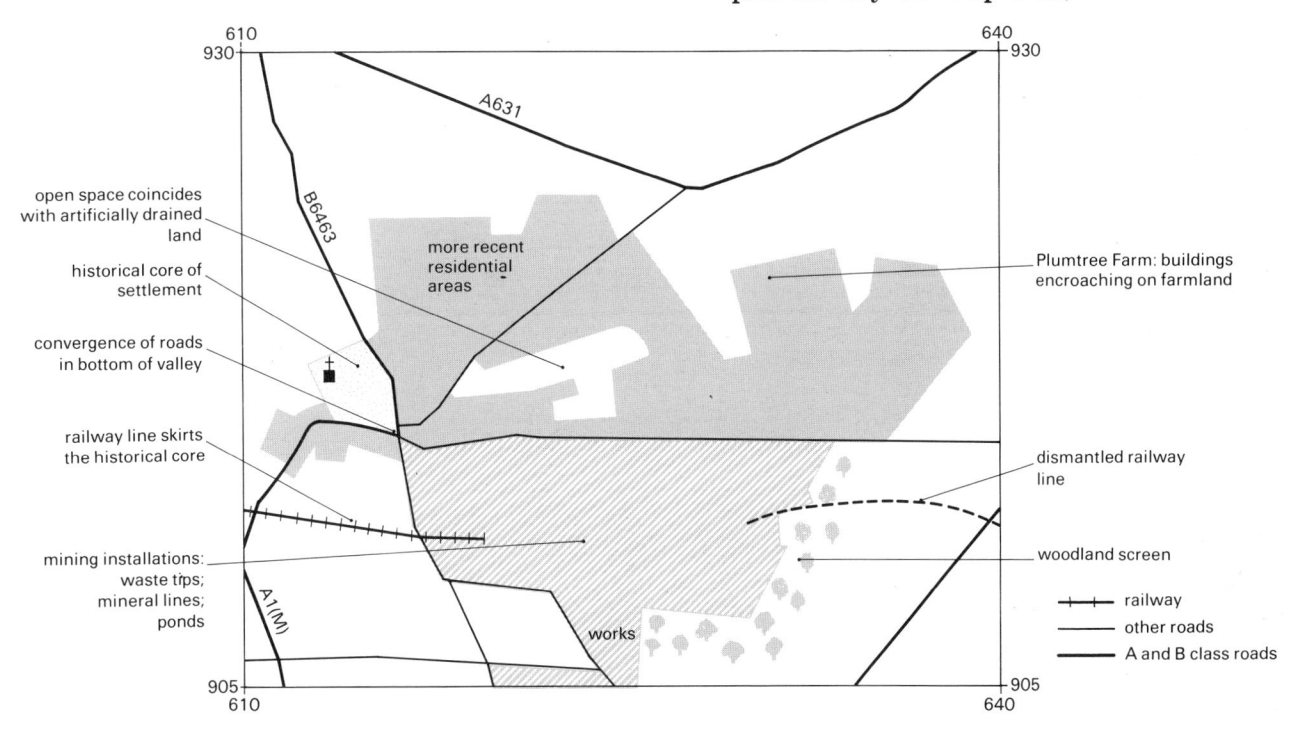

Fig. 9.7 Major land-use regions in Harworth

(ii) separated from the historical core is a clearly defined industrial zone, partly screened by tree planting and serviced by mineral lines;

(iii) a recent belt of residential development, encroaching on the countryside, and comprising housing estates with services e.g. schools, post offices, churches.

Are the same land-use regions represented in other nucleated settlements? If so, we are justified in describing these settlements, too, as mining centres.

9.29 Examine New Rossington/Rossington between Eastings 595 and 640, and complete the following tasks.

(i) Prepare a sketch map to show: (a) the perimeter (edge) of the built-up area, (b) the main roads and railways, (c) the following land-use zones — mining area, historical core, new residential area, woodland/forestry.

(ii) Compare your sketch map with Fig. 9.7. Do you think that the pattern of land-use regions in New Rossington/Rossington compares with that for Harworth? Give reasons to support your answer.

9.30 Which settlements shown on the map would you classify as mining settlements? Give reasons for your choice, and illustrate your answers with suitable sketch maps.

Index

accessibility 17, 65, 95
  to services 71–2
agriculture 17, 20, 38–9, 45–6, 85, 95–6, 104
air photographs 12–14
  interpretation 22, 36, 40, 44–5, 56–7, 60, 69–70, 71, 77, 84, 109
  oblique 13
  vertical 13–14
air pollution 70
alluvial fan 42
area
  of drainage basin 29–30
  enlargement 7, 27
  measurement 30
  reduction 7
arête 33
artificial drainage *see* drainage
aspect 17, 20
associated number 65
average gradient 31

bay 57
bay bar 57
barrier lake 57
bridging point 69, 78, 81, 107
building density 79, 80, 85, 93
built-up area 85, 86, 93, 97, 107; *see also* settlement analysis; settlement
by-pass 83, 95, 98–9,
  model 98

canal 60, 70
car park 71, 79
carr 102
Central Business District (CBD)
  characterstics 70, 71, 78–80
  expansion 71, 80
chalk 8, 15, 17
clay 9, 16, 19
  vale 15, 16, 17
cliff 57, 58
coastal landforms 48–59
coastline
  emergent 55
  submergent 55
communications 17, 20, 60–66, 70, 71, 76, 77–8, 93, 104
commuter belt 79
compass direction 2
connectivity 63–4
contour 7–8
  definition of 7
  spacing 11
contour trace 9, 48, 49, 67, 74, 85, 101
corrie 32–3, 34, 35

deposition 48, 57–8
detour index 66, 72
dip slope 8, 10, 12, 16, 18
direction 2
dissection 14, 50
drainage
  artificial 17, 73, 91, 102
  channel 20
  density 30–31
  human interference 20, 46, 68, 73–5, 102
  network 29, 101
  pattern 14, 17, 21, 29, 91
  systems 28–9, 67, 73, 85, 91
  urban 73–5, 91–2
drainage basin 20, 29–31
dry valley 8, 10, 12, 16
dyke 75

enlargement 7
erosion
  glacial 33, 34
  marine 48, 57, 58
  river 21, 22, 59
estuary 55–6, 59, 75, 78
extractive industry *see* mining; quarrying

farm 18, 19, 59, 61, 95, 96, 104, 107, 108
farming *see* agriculture
field boundaries 45
flood protection 45–6, 68, 71, 75, 91; *see also* drainage, artificial
floodplain 21, 27, 28, 68, 73, 102
forestry 39; *see also* woodland
Forestry Commission 39, 96, 97, 104

geology 8–9, 42–4
Georgian architecture 70
glaciation 32–8
golf course 18, 83, 94, 96, 97
gradient 7, 20, 34, 68, 102
  average 31
  measurement 11–12
grid-iron 80, 82, 83
Green Belt 95, 96

hachures 7
hanging valley 34–5, 36
headland 57
headwall 33
high water mark 55, 74, 75
hill fort 27
historical maps 68–9
horticulture 18, 95–6
house type 47, 70, 72, 80, 81, 84, 109

housing estate 81, 83–4, 94, 95, 110

igneous rocks 9
industrial estates 76, 77
industry
  buildings 75–7
  communications 20, 60, 76, 77–8
  and drainage 73
  land use 70, 77, 82, 83, 85, 102, 103, 109–10
  oil refining 75, 76
  regions 70, 77–8, 81, 85
  site 78
  see also mining; quarrying; tourism

interlocking spurs 21, 28, 50
intervisibility 7
island 57
isopleth 65

joint patterns 8
journey *see* accessibility
  to shop 80, 110
  to school 47, 72, 80, 110
  to work 83, 100

Konig Number 65

land use *see* agriculture; recreational land use; industry; settlement
land-use model 79
land-use transect 93
levées 21, 28
limestone 8–9, 20, 42–4, 101
line section 10–11
  annotation 10, 11, 45, 50, 73, 74, 102
  construction 11, 42, 47, 67, 68
  longitudinal 21
  sketch 11
longshore drift 57–8

map
  distance 2
  location 32
  orientation 2
  references 1–2
  sketch 14–15, 37, 105
marine
  deposition 48, 57–8
  erosion 48, 57, 58
  platforms 55
marshland 28, 50, 102
  coastal 73, 75, 85
meander 21, 27, 28, 35
metamorphic rocks 9

metropolitan fringe 94
Millstone Grit 9, 42, 43, 46
mineral line 102, 103, 104
mining 46, 47, 102, 103, 109
   disused mine 46
motorway 85, 99–100, 104

national grid 1, 3
National Trust 96, 97
nature trail 40
network analysis 62–6

oolitic limestone 70
open space 85, 94, 95, 107, 109
orchards 96
orientation 2
outlier 15, 22
ox-bow lake 21, 28, 19

park 70, 83, 93, 95, 97
plateau 48–50
port 59
pot hole 9, 42
public buildings 70, 71, 93–4
public services 47, 71–2, 81, 83, 94
public utilities 94, 95
pyramidal peak 33–4

quarrying 39, 40, 46, 60, 76, 102, 103

railway 60, 61, 70, 72, 78, 95, 98, 100, 104
   disused 29, 59, 60, 61, 103, 104, 109
raised beach 55
recreational land use 18, 27, 40, 46, 59, 96–7
reduction 7
region, delimitation 14–16, 36–9, 48–50, 101–2; *see also* industry; settlement
relative relief 12, 41–2, 50
reservoir 20, 31
residential area 59, 60, 70, 72, 79, 80, 84, 109, 110
residential environment 60, 72, 84
ria 55, 59
river 20–31
   basin 29–31
   capture 28, 29, 50
   cliff 22, 27
   course: description 28; upper 21; middle 21–2, 27; lower 21, 28
   length 31
road network 60–6, 95, 98–100
rock types 8–9, 42
route
   description 61–2, 98–9
   efficiency 66
   network 62–6, 97–100, 102, 104
rural settlement *see* settlement; village

sands, 9, 16, 19
sandstone 9
scale 1, 2, 13, 80
scar 7, 8, 9, 10, 42

scarp 8, 9, 10, 15–16
scarpland 10, 16, 17, 22
scree 7, 34
sea level
   datum 7
   change 55
settlement
   classification 107
   dispersion 105
   economy 60, 68, 107
   historical evolution 16, 18, 46, 68, 71, 80–1, 108–10
   nucleation 47, 105
   pattern 16, 40, 47, 60, 81, 82–4, 104–5
   planning 79, 80, 95, 96
   regions 16
   rural 60, 104–6; *see also* village
   size 105, 107
   spring-line 16, 18–19
   urban fringe 94–7
   urban sprawl 83, 85
settlement analysis
   form (morphology) 46–7, 107; *see also* street pattern, grid-iron
   function 47, 107
   site 19, 20, 60, 69, 105
   situation 105; *see also* residential environment
shale 42
Shimbel Index 65
sketch
   section 11
   map (annotated) 14–15, 37, 105
slope
   concave 11, 16
   convex 11, 16
   measurement 11, 12
   profile 11, 12
spa 70
spoil tips 104, 109
spot height 8, 9, 68, 73
spring line 10, 18
spur 8
stack 57
street patterns 72, 80–84, 94
surface water 20–21; *see also* drainage
superficial deposits 9
symbols
   for 1: 50 000 maps 4, 5
   for 1: 25 000 maps 6
   for 1: 10 000 maps 6
   for sketch maps 14

tombolo 57
topology 62–6
tourism 20, 40, 46, 59, 60, 70
town walls 70
townscape 70–71
transect diagram 44–5, 93, 94, 102, 103, 108
trend line 48–50
triangulation pillar 8, 9, 73
truncated spur 57, 58

U-shaped valley 34–5
urban settlement *see* settlement

analysis; settlement; residential environment

valley form 21–8, 44, 73–5, 85, 91
vegetation 16, 17, 20, 45; *see also* agriculture; woodland
villages
   classification 107
   dormitory 82
   industrial 40
   mining 40, 109–10
   satellite 95
   types 94, 104
   see also settlement anaylsis; settlement

waterfall 28, 35, 38
watershed 15, 29–30, 37, 38
woodland 17, 19, 39, 45, 57, 93, 96, 97, 104, 105

Yoredale Series 42, 43